Progress in Theoretical Computer Science

Roberto Di Cosmo

Isomorphisms
of Types:
from λ-calculus
to information retrieval
and language design

Birkhäuser   1995
Boston • Basel • Berlin

Roberto Di Cosmo
LIENS-DMI
Ecole Normale Supérieure
75230 Paris Cedex 05
France

Library of Congress Cataloging-in-Publication Data

Di Cosmo, Roberto, 1963-
    Isomorphisms of types : from [lambda]-calculus to information
retrieval and language design / Roberto Di Cosmo.
        p.   cm. -- (Progress in theoretical computer science)
    Includes bibliographical references and index.
    ISBN-13: 978-1-4612-7585-5
    1. Programming languages (Electronic computers).   2. Type theory.
3. Human-computer interaction.   I. Title.   II. Series.
QA76.7.D5   1993   94-36650
005.13'1--dc20   CIP

Printed on acid-free paper

Birkhäuser ®

© Birkhäuser Boston 1995

Softcover reprint of the hardcover 1st edition 1995

ISBN-13: 978-1-4612-7585-5   e-ISBN-13: 978-1-4612-2572-0
DOI: 10.1007/978-1-4612-2572-0

Typeset by the author.

9 8 7 6 5 4 3 2 1

Contents

1	**Introduction**		**11**
	1.1	What is a type?	12
	1.2	Types in mathematical logic	13
	1.3	Types for programming	14
		1.3.1 Imperative languages	15
		1.3.2 Limits of static type-checking	19
		1.3.3 Functional languages	20
		1.3.4 The lambda calculus	21
	1.4	Exploring typed λ-calculi	23
		1.4.1 Church-style types	23
		1.4.2 Curry-style types	24
		1.4.3 Explicit polymorphic types	24
		1.4.4 Implicit polymorphic types	25
⚠	1.5	The typed λ-calculi used in this work	27
		1.5.1 The calculus $\lambda^2\beta\eta\pi*$	27
		1.5.2 General notations for terms and substitutions	28
	1.6	The Curry-Howard Isomorphism	30
	1.7	Using types to classify and retrieve software	34
		1.7.1 Object-oriented languages	35
		1.7.2 Functional languages	36
	1.8	When are two types equal?	37
		1.8.1 Isomorphic types	38
		1.8.2 Isomorphisms in category theory	41
⚠		1.8.3 Digression: Tarski's High School Algebra Problem	43
		1.8.4 Isomorphisms in logic	45
	1.9	Isomorphisms and the lambda calculus	46
⚠		1.9.1 Isomorphisms and invertibility	46
		1.9.2 The theories of isomorphisms for typed λ-calculi	49
⚠		1.9.3 Soundness	50

2 Confluence Results   **57**

2.1   Introduction   58
2.2   Extensionality   61
    2.2.1   Survey   62
2.3   Overview   67
    2.3.1   Weakly confluent reduction   67
    2.3.2   Investigating strong normalization   71
    2.3.3   A general criterion for confluence   72
2.4   Confluence   74
2.5   Weak normalization   79
2.6   Decidability and conservative extension results   81
2.7   Other related works   81

3 Strong normalization results   **85**

3.1   Normalization without β^2 on *gentop* n.f.'s   86
    3.1.1   Reducibility with parameters   89
3.2   Normalization without η_{top} and SP_{top}   96

4 First-Order Isomorphic Types   **101**

4.1   Rewriting types 103
4.2   From $\lambda^1\beta\eta\pi*$ to the classical $\lambda^1\beta\eta$ 105
4.3   Using finite hereditary permutations 112
4.4   The complete theories of $\lambda^1\beta\eta\pi$ and $\lambda^1\beta\eta*$ 116

5 Second-Order Isomorphic Types   **119**

5.1   Towards completeness 120
    5.1.1   Outline of the section 120
    5.1.2   Reduction to a subclass of types 121
    5.1.3   Reduction to a subclass of terms 123
5.2   Characterizing canonical terms 124
    5.2.1   Outline of the section 124
    5.2.2   Projection of invertibility over coordinates 125
    5.2.3   Reduction of coordinates to $\lambda^2\beta\eta$ 130
    5.2.4   Syntactic characterization of canonical bijections . . 137
5.3   Completeness for isomorphisms 139
    5.3.1   Uniform isomorphisms 143
5.4   Decidability of the equational theory 143
5.5   The complete theories of $\lambda^2\beta\eta\pi$ and $\lambda^2\beta\eta*$ 145
5.6   Conclusions 146
A   Properties of n-tuples 147
B   Technical lemmas 151
C   Miscellanea 161

6   Isomorphisms for ML   **165**
  6.1   Introduction 166
  6.2   Isomorphisms of types in ML-style languages 167
      6.2.1   A formal setting for valid isomorphisms in ML-like languages 168
  6.3   Completeness and conservativity results 172
      6.3.1   Completeness 174
      6.3.2   Relating $Th^2_{\times T}$ and Th^{ML} 174
  6.4   Deciding ML isomorphism 175
      6.4.1   An improved decision procedure 177
      6.4.2   Equality as unification with variable renamings . . . 179
      6.4.3   Dynamic programming 180
      6.4.4   Experimental results 185
  6.5   Adding isomorphisms to the ML type-checker 188
      6.5.1   Type-inference with just (**Split**) 191
      6.5.2   What is special in (**Split**) 191
      6.5.3   Choosing the right isomorphisms 193
      6.5.4   Right isomorphisms in impure context 195
  6.6   Conclusion 195
  6.7   Some technical Lemmas 196
  6.8   Completeness 196
  6.9   Conservativity 200

7   Related Works, Future Perspectives   **205**
  7.1   Equational matching of types 206
      7.1.1   Decomposing the matching problem 207
  7.2   Using equational unification 210
  7.3   Extending the paradigm 210
      7.3.1   Searching through type classes 210
      7.3.2   Searching with more powerful specifications 211
      7.3.3   Recursive terms and types 211
      7.3.4   Other applications of type isomorphisms 212
  7.4   Future work and perspectives 212
      7.4.1   Design of type systems for functional languages . . . 212
      7.4.2   Object retrieval in object-oriented libraries 213
      7.4.3   Dynamic composition of software components 213
      7.4.4   Representation optimization 213

Bibliography   215
Subject Index   233
Citation Index   237

Isomorphisms of Types: From λ-calculus to Information Retrieval and Language Design

a Bruna e Dario,
con affetto

Acknowledgments

This book is the outcome of several years of interaction with many persons without whom this work would not have seen the light, but in all these years some very special persons have been constantly at my side, and I would like to begin by thanking them. My parents and all my family, who know that there are no words to express my gratitude to them. I had the chance to meet Delia Kesner and to share with her the wonderful experience of building day after day a strong relationship that goes far beyond our scientific cooperation.

I'm greatly indebted to my advisor, Giuseppe Longo, for his continuous encouragement, discussions and insights: he strongly motivated me to carry on all the work that supports the results presented in this book.

I met Pierre-Louis Curien on my arrival in Paris, and I enjoyed with him the wonderful experience of the investigation of $\lambda^2\beta\eta\pi*$, the core of Chapters 2 and 3 of this book. He knows how much I owe him.

Thanks to Roger Hindley and Jerzy Tiuryn who made me the honor of being referees for this book.

I would like to express my gratitude to Mikael Rittri, whose work on library searches provided interesting practical applications to the theoretical results developed in all this work: he has been a constant and copious source of information on the ongoing work in that area. Thanks go also to Gregory Mints for pointing out to us the work of Sergei Soloviev: I met Sergei shortly after I discovered his works, and I would like to thank him for the many fruitful and interesting discussions we had since then.

I wish to thank Hubert Comon and Jean-Pierre Jouannaud, as well as all the working group of the LRI at the University of Orsay, for many stimulating discussions: they helped and pushed me in many essential phases.

The whole group working at INRIA at the CAML system has provided invaluable support in the development of the code that implements the library search system, and I'm particularly grateful to Pierre Weis, whose help was crucial in the integration of this new tool in the CAML system, and to Xavier Leroy, whose impressive programming skill led to a working experimental implementation of the new type-inference mechanism for CamlLight in less than an afternoon.

The Liens and the Dmi, at the Ecole Normale Supérieure in Paris, helped my work by providing a fantastic scientific environment and excellent facilities.

A special thank goes to Robert Constable, for hosting me in Cornell's unique environment for six unforgettable months.

Thanks finally to Franco Barbanera, Emmanuel Chailloux, Pierre Cregut, Mariangiola Dezani-Ciancaglini, Vincent Danos, Furio Honsell, Martin Hyland, Eugenio Moggi, Chetan Murthy, Paliath Narendran, Adolfo Piperno, Andrew Pitts, Laurent Regnier, Simona Ronchi-Della Rocca, Pino Rosolini, André Scedrov, Val Tannen and many others for several interesting discussions on all these matters.

Preface

This is a book about *isomorphisms of types*, a recent difficult research topic in type theory that turned out to be able to have valuable practical applications both for programming language design and for more human-centered information retrieval in software libraries. By means of a deep study of the syntax of the now widely known typed λ-calculus, it is possible to identify some simple equations between types that on one hand allow to improve the design of the ML language, and on the other hand provide the basis for building radically new information retrieval systems for functional software libraries.

We present in this book both the theoretical aspects of these researches and a fully functional implementation of some of their applications in such a way to provide interesting material both for the theoretician looking for proofs and for the practitioner interested in implementation details.

In order to make it possible for these different types of readers to use this book effectively, some special signs are used to designate material that is particularly technical or applied or that represents a digression. When the symbol

appears at the beginning of a section or a subsection, it warns that the material contained in such section is particularly technical with respect to the general level of the chapter or section where it is located. This material is generally reserved to theoreticians and does not need to be read by the casual reader.

The symbol

marks instead those sections, or subsections, that constitute a "detour" in the train of thought. Such material is generally of interest on its own and allows a better understanding of the subject, but it does not need to be read immediately.

Finally, the symbol

marks sections that contain material particularly useful for applied readers interested in implementing new systems based on the theoretical results that they probably do not want to study fully.

Let we just briefly remark that the first chapters contain mostly theoretical sections, while the applied material is mostly to be found in the last Chapters.

Source code

Implementations of the search tools described in this book are available for many different functional programming languages.

Mikael Rittri's retrieval system for Lazy ML is available on request (e-mail: `rittri@cs.chalmers.se`). One version uses unification modulo isomorphisms.

Brian Matthews's retrieval system for Haskell is available on request (e-mail: `bmm@inf.rl.ac.uk`). This system handles Haskell's type-classes.

The Venari project's search tool for SML, with an interface in Gnu Emacs, is available on request from `amy+@cs.cmu.edu`.

Finally, the source code for the complete implementations of the library search systems described in Chapter 6 is available by **anonymous ftp** from `ftp.inria.fr` in

`/lang/caml/V3.1/common.tar.Z` : look at files in `user_lib/FIND_IN_LIB` (equality modulo isomorphisms for CAML), and

`/lang/caml-light/cl6unix.tar.Z` : look at files in `contrib/search_isos` (equality and matching modulo isomorphisms, with NeXT and Gnu Emacs interface, for CamlLight).

Julien Jalon and Jerôme Vouillon implemented during their *Projet d'informatique* the search tools based on matching.

Notation index

Notation	comment	page
$f[g/x]$	substitution of g for x in f	22
$\Gamma \vdash M : A$	M has type A under assumptions Γ	23
$\lambda^2\beta\eta\pi*$	second-order λ-calculus with pairs and unit type	27
$\mathbf{T}$	terminal object in a category	27
$*$	unique element of type $\mathbf{T}$	27
$=_{\beta^2\eta^2\pi*}$	equality for $\lambda^2\beta\eta\pi*$	28
$[M_1,\cdots,M_n]$	sequence of terms	29
$\vec{M}$	sequence of terms	29
$\|\vec{M}\|$	length of sequence $\vec{M}$	29
I_A	$\lambda x : A.x$, the identity of type A	29
$N[\vec{M}/\vec{x}]$	simultaneous substitution of $\vec{M}$ for $\vec{x}$	29
$FV(M)$	free variables of M	29
$FTV(A)$	free type variables of A	29
$\langle M_1,\cdots,M_n \rangle$	n-tuple	29
$\lambda^2\beta\eta*$	second-order λ-calculus with pairs	29
$=_{\beta^2\eta^2\pi}$	equality for $\lambda^2\beta\eta\pi$	29
$\lambda^2\beta\eta*$	second-order λ-calculus with unit type	30
$=_{\beta^2\eta^2*}$	equality for $\lambda^2\beta\eta*$	30

Notation	comment	page
$\lambda^1\beta\eta\pi*$	first-order λ-calculus with pairs and unit type	30
$=_{\beta\eta\pi*}$	equality for $\lambda^1\beta\eta\pi*$	30
$\lambda^1\beta\eta\pi$	first-order λ-calculus with pairs	30
$=_{\beta\eta\pi}$	equality for $\lambda^1\beta\eta\pi$	30
$\lambda^1\beta\eta*$	first-order λ-calculus with unit type	30
$=_{\beta\eta*}$	equality for $\lambda^1\beta\eta*$	30
$A \cong_d B$	A and B are definably isomorphic	39
$M \circ N$	composition $\lambda x.\lambda f.\lambda g.(f(gx))$ of λ-terms	39
$A \cong B$	$\mathcal{M} \models A \cong B$ for all $\mathcal{M}$	39
$\mathcal{M} \models A \cong B$	A and B are isomorphic in the model $\mathcal{M}$	39
IPC	intuitionistic positive calculus	45
BT(M)	Böhm tree	47
Th^1	theory of isomorphisms of $\lambda^1\beta\eta$	52
$Th^1_\times$	theory of isomorphisms of $\lambda^1\beta\eta\pi$	52
Th^1_T	theory of isomorphisms of $\lambda^1\beta\eta*$	52
$Th^1_{\times T}$	theory of isomorphisms of $\lambda^1\beta\eta\pi*$	52
Th^2	theory of isomorphisms of $\lambda^2\beta\eta$	52
$Th^2_\times$	theory of isomorphisms of $\lambda^2\beta\eta\pi$	52
Th^2_T	theory of isomorphisms of $\lambda^2\beta\eta*$	52
$Th^2_{\times T}$	theory of isomorphisms of $\lambda^2\beta\eta\pi*$	52
Th^{ML}	theory of isomorphisms of core ML	52
n.f.	normal form	58
$\rightarrow$	one-step reduction	58
$\rightarrow_=$	one- or zero-step reduction	58
$\overset{*}{\rightarrow}$	zero or more step reduction	58

Notation	comment	
$\xrightarrow{R}$	one-step R-reduction	58
$\xrightarrow{R}_=$	one- or zero-step R-reduction	58
$\xrightarrow{R}{}_*$	zero or more step R-reduction	58
CR	Church-Rosser Property	59
WCR	weak Church-Rosser Property	59
SN	strong normalization	59
WN	weak normalization	59
η	extensionality	62
SP	surjective pairing	62
$Iso(\mathbf{T})$	types isomorphic to $\mathbf{T}$	69
$\xrightarrow{\beta^2\eta^2\pi*}$	reduction for $\lambda^2\beta\eta\pi*$	69
$gentop$	generalized Top reduction	69
SP_{top}	generalized SP reduction	69
η_{top}	generalized η reduction	69
$(M)^T$	$gentop$ normal form of M	74
$\overset{gentop}{\xrightarrow{\quad*\quad}}$	reduction to $gentop$ n.f.	74
T°	terminal-free type isomorphic to T	82
$\nu(u)$	legth of the longest reduction path of u	87
$RED_T[\overrightarrow{R}/\overrightarrow{X}]$	reducibility with parameters	89
$\mathcal{R}^1$	first-order type rewriting	103
$\mathcal{R}^2$	second-order type rewriting	121
$[\overrightarrow{\mathrm{p}_i(Nw)}/\overrightarrow{x}]$	compact notation for $[\mathrm{p}_1(Nw),\cdots,\mathrm{p}_m(Nw)/x_1,\cdots,x_m]$	127
$(P\{M[N/x]\})$	(PM) where $[N/x]$ is applied only to M	131
$\mathrm{p}_i^k M$	ith projection for a k-tuple	147
Th^{ML}	theory of isomorphisms for ML	174

Notation	comment	page
$s.n.f.(A)$	split normal form of type A	175
pts	principal type schema	193

Chapter 1

Introduction

In modern programming languages the notion of type has become so natural and widespread that it is easy to forget about its origins and its theoretical relevance. A *type* is simply seen as a useful means of classification of the objects that a program manipulates: types help to understand better what a program does, and they also provide a valuable firewall against many common programming errors.

But the theory of types is also a well-established and very active field of research in mathematical logic and theoretical computer science, one which can often generate interesting practical fallout. In this book we give a detailed account of one such successful story of nice theory that produced valuable practical fallout: the study of *isomorphic* types, originally started from very theoretical motivations in category theory and in the theory of λ-calculus, turned out to provide a firm basis both for the design of new programming languages and for the development of natural tools for information retrieval in software libraries.

The first sections up to 1.6 of this introduction provide a gentle and general introduction to the notion of type. We start from its origins in mathematics, and then we focus on its modern use in programming languages and type theory, showing how the typed λ-calculus can be of great help to understand many key features of the type systems available in modern programming languages.

Finally, in Section 1.7, we give a general overview of the reasons why types are good candidates to classify software components and how they can be used as retrieval tools, with examples and motivations coming from different programming languages. This will bring up the question of when two types contain the same information and must then be identified; this is addressed in the Section 1.8, with broader references to proof theory, category theory and lambda calculus.

1.1   What is a type?

The modern notion of *type* makes its first appearance in the field of mathematical logic, where it was introduced long before the appearance of digital computers and modern programming languages. As we will briefly recall, two different approaches to the foundation of mathematical logic - one based on sets and the other on functions - had to resort to a notion of *type* in order to avoid paradoxes, and both were soon abandoned as logical formalisms. But the notion of type survived and found a fundamental place in computer science, where it is now an essential tool to design usable programming languages, to prove properties of programs, to ensure data encapsulation and hiding, and to retrieve software components as well as toward many more applications.

We will now first recall how types developed from set theory, then look at how they evolved in imperative programming languages, and finally follow the smooth evolution of types in the theory of the λ-calculus, which showed up as an essential connection tool between logic and functional programming.

1.2   Types in mathematical logic

At the end of the 19th century, Frege made a revolutionary effort towards a global formalization of mathematics. It is contained in many now famous works (like [Fre79, Fre93, Fre03]), where a first formal theory of set is presented as a basis for the formalization of all the mathematical activity.

In Frege's system, sets are taken as the basic building blocks for a complete formalization of mathematics and are very liberally built out of logical formulae by means of what in modern terms is called the axiom of comprehension. In very simple terms, this means that whenever you are able in Frege's formalism to write down a particular property (like, for example, the property of being an even number), then you can immediately form the set of all objects having such a property (in the example, the set of all even numbers).

Such a liberal system, as discovered in 1902 by Russel, allows us to define paradoxical objects like the set of the sets that are not elements of themselves, $R = \{x | x \notin x\}$ in modern terms. Indeed, "not being an element of itself" is a property that can be formalized in Frege's system, and then R must be a set. But then, is R an element of itself? Either answer to this question is in contradiction with the definition of R, and this fact goes now under the name of Russel's paradox. Due to this inconsistency, Frege's original system allows to prove any proposition, and so has no logical interest.

In order to avoid the paradoxes of Frege's system, Russel and Whitehead introduced in [RW10] a classification of the sets using *types*: a kind of tag that is attached to every set. By using these types, they could properly restrict the set-forming operations so that, for example, one is no longer able to build a set like R. The price to pay is an explicit classification of the sets. This can be considered as the origin of the notion of type, but it was just the beginning of a long and fruitful line of research that shifted over the years from the foundation of mathematics to theoretical computer science, where is now best studied. The interested reader will find a nice historical survey of types in mathematical logic in [CF58, CHS72], while we will focus now on the use of types in computer science.

1.3   Types for programming

The evolution of the notion of type for programming languages followed a somewhat similar path. The first programming languages had no notion of type at all, and there was no explicit classification of the data, which were all uniformly represented as memory words. This gave a very simple framework for writing programs and an extraordinary freedom to the programmer.

When writing programs, though, no matter the programming language we use, we start immediately to partition the universe of the data we handle into different groups of objects: lists, integers, characters, reals, vectors and so many other classes of beings that we wish to consider as objects of different kinds, and that we intend to use for different purposes.

This does not automatically force us to select a typed programming language. After all, the main memory of a traditional computer is almost a perfectly untyped universe, where different types are suitably coded in terms of the underlying uniform memory words. In the first programming languages, there was no real notion of *type of data*: In Lisp all the data are S-expressions, while in Fortran there is only a very rudimentary distinction between integers and real numbers, coming more from efficiency considerations than other reasons.

But a language without types allows us to write a lot of programs that perform operations generally understood as unsafe or nonsense, like extracting the square root of (the code of) a character, or trying to execute (the code of) an array of real numbers as if it were (the code of) a program, something that is the programmer's analog of a logical paradox. Unfortunately, these kinds of nonsense operations are not necessarily the result of a program written by some brain-damaged hacker, but they can show up as the result of rather subtle (and very difficult to trace down) logical errors: any experienced Fortran user surely faced such a situation when using typical type-unsafe features of the language as the `common` or `equivalence` statements.

A way to ensure the consistency of the operations is to associate a *type* to every data object and to devise a suitable set of rules that use these types to decide when it is safe to perform a given operation on a given set of objects. For example, given a function $f : A \to B$, and an object o of type A, we can say that it is legal to apply f to o and get fo of type B. These rules are usually stated in the form of deduction rules as in

$$\frac{f : A \to B \qquad o : A}{fo : B} \ .$$

Several programming languages, both imperative and declarative, have addressed this need for a formal mechanism that take care, at least to some extent, that operations on the data are carried on in a consistent way. The solutions proposed all agree in using some notion of *type* that is attached to the objects manipulated by the program and that is checked in order to prevent type errors. Anyway, they may differ in several respects. Let us review here some of the most relevant ones.

When the types are checked: the verification of the consistency of the operations can be performed at compile-time (*static* type-checking) or deferred to run-time (*dynamic* type-checking).

Who is required to fill in the type information: the types can be inferred by the compiler (*implicit typing* or *type-inference*) or they must be declared explicitly by the programmer (*explicit typing*).

Safety: the type system guarantees that there will be no run-time type errors (*strong typing*) or just rule out some restricted class of type errors (*weak typing*). A language with a weak type system is usually compiled with extra code added to watch for impending type errors, so a statically typed language can run usually faster than a dynamically typed one.

Flexibility: a type system can often refuse as erroneous those programs that will not generate type errors at run-time, simply because the system is not powerful enough to discover that this is the case. Type systems can have different degrees of flexibility, ranging from basic *monomorphic* systems that allow any piece of code to be used only with one specific type [1] to highly complex *polymorphic* systems that allow to reuse the same piece of code at different types.

In general, we will clearly prefer a static, strong, flexible, implicit type system. Typing also greatly improves readability of a program by clearly classifying data and by providing a partial specification of the behavior of a function. If we know that a function takes two integers and returns another one, we can read its code with more confidence than when just knowing that it takes two arguments.

1.3.1 Imperative languages

Along with the evolution of imperative languages, we can follow closely an evolution of the type systems technology. In the first high-level imperative language Fortran there was a very weak type system that just gave the user

---

[1] This imposes to rewrite over and over again the same code. A typical example is the **sort** function in Pascal: you must write down one version of it for every type of object you want to sort.

only real numbers, integers and arrays (no strings!), while providing very simple and powerful means to break even this rudimentary type discipline. Using the `common` and `equivalence` constructs, it is straightforward to write the representation of a real number into some memory locations and then force the program to read later these very locations as if they contained the representation of an integer, with catastrophic consequences for the computation.

These kind of errors are extremely difficult to trace down in the absence of a strong typing discipline: not only does the compiler not produce warnings for the user, but also, due to specific implementation details, an erroneous program can happen to work properly if compiled on one machine, and break down mysteriously when recompiled on a different system.[2]

This was a strong motivation to design better type systems for later languages: in Pascal ([JW72]), for example, we find a clear example of a strong type system. The user is provided four base types (reals, integers, characters and booleans), and has a set of tools to build his own more complex (even recursive) types using (fixed length) arrays, records and a safe version of pointers. The type system of Pascal is strong and explicit: the programmer must explicitly declare the type of each identifier before using it, and once an identifier is declared, say, of type real, then it will not be possible for it to contain a value of some other type during execution. This guarantee against run-time type errors comes at a price, though, in Pascal: the type system is also monomorphic, and this is bad news for code reusability. Imagine you want to write a procedure sort that takes an array of objects, a comparison function, and gives you back the same array where the objects are properly ordered. The sorting algorithm does not depend on the type of the objects to be sorted, so that it would be desirable to write down a generic sorting procedure that will work on any input array, regardless of its type. Unfortunately, in Pascal you are obliged to declare the type of each identifier (including the formal parameters of the procedure), and then you must declare the type (and only one type) of the array parameter; this is what a typical Pascal declaration for a sort procedure would look like:

```
procedure sort
  (function comp(x: t, x: t): boolean, v: array [1..300] of t)
```

We are forced to write one copy of the code to sort integers, another

---

[2]This typically occurs with reals or integers of different *lengths*. Imagine that your compiler happens to code a long real that occupies twice the space of a short real, with its least significant part first; then an erroneous identification of a long real with a short real in a `common` may not produce an error, but will probably break the program if you use a different compiler.

to sort reals, and so on, even if the only difference in the code is the type declarations for the formal parameters of the procedure.[3]

As far as the type systems go, there were essentially two main reactions to the limitations of Pascal's monomorphic system: go for a more powerful type mechanism that allows to reuse the source code (and this was the choice of ADA), or provide the programmer with powerful tools to break the type system, if necessary (and this was the choice of C).

In ADA ([Ame83]) there is the notion of **generic** that allows to write a program parameterized with respect to some data structures that it does not need to know in details[4]: you can write a generic sorting routine **sort** parameterized w.r.t. the type t of the objects of the array to be sorted:

```
generic
    type typ is private
    with function comp(x: typ, y:typ)
procedure sort(a: array (1..MAX) of typ)
```

Then, in order to obtain an integer sorting routine, you just take an instance of the generic program by declaring that **typ** is now **int**:

```
procedure int_sort is new sort(int,"<")
```

Notice that in ADA this mechanism *does* produce a new copy of the code for each instance, but it is no longer the programmer who is obliged to write it down.

In C ([KR78]) the notion of safe pointer present in Pascal is heavily modified by allowing to have pointers to, and to read the address of, practically any structure of the language (including functions), using the * and & operators. This allows programmers to write in a very easy way a sort function that can work on any structure (and without creation of new copies of the same code). This actually comes out of the fact that using C pointers one can pass around any data structure using just one type of object, the pointer itself, seen as an address into physical memory. This mechanism is so practical that it is very widespread. Here is, for example, the commonplace definition of the quicksort routine from a Unix standard library:

```
void *qsort(void *base, int nmemb, int size, int (*compar)());
```

---

[3] Actually, due to the fact that Pascal arrays are indeed static, the 300 in the declaration is mandatory, so you will need a different version of the procedure to sort 1000 elements!

[4] This facility is really motivated by concerns for data abstraction and encapsulation, but it also provides a fair degree of polymorphism.

The first argument is a pointer to the base of the data; the second is the number of elements; the third is the size of an element in bytes; the last is the name of the comparison routine to be called with two arguments which are pointers to the elements being compared. This is an example of application, taken from a Unix manual page:

```
static    int charcompare(i,j)
char *i, *j;
{
    return(*i - *j);
}


main()
{
    char a[10];
    int i;
    a[0] = '9';
    a[1] = '8';
    a[2] = '7';

    qsort(a,3,sizeof(char),charcompare);
    for (i=0; i<3; i++) printf(" %c",a[i]);
    printf("\n");
}
```

By looking carefully at this *perfectly legal* C code, it is easy to understand what is going on. In C we have enough tools to see the real machine behind the programming language, and we can go back to the days of the assembler languages, where we could do anything with the data in memory. Here, for example, we declare that our formal function parameter compar takes as arguments two pointers to a fake data structure void that is used just to fool the weak C type system, while what we really want from these pointers is just the real memory addresses of the objects to compare. In the example, the actual parameter takes as arguments two character pointers. Another *feature* we use here is the automatic coercion of characters to integers that make the code of charcompare so easy to write and so surprising for beginners: we are subtracting (the code of) two characters! Last, but not least, the definition of the formal parameter compar of the qsort function is surprising too: it is declared as a function of zero arguments, but is really used with two arguments. Indeed, in many C compilers you can pass to a function as many arguments as you like, less or more than what is necessary: it is up to you to know what you are doing. Not apparent in this example but even more disturbing is the possibility of breaking the scope of local variables: by using the address operator &, you can get the address of a local variable and pass it to another function!

This programming scheme is clearly very flexible, and one can write simple functions that seem quite polymorphic, but all these "facilities" correspond to different ways of breaking type rules: we are really in the presence of an almost type-free programming language. The result is often a bad and obscure programming style that produces systematically potential type errors that the C compiler is unable to spot.[5]

Such extremely weak typing rules in C lead to so much trouble, that in successive languages, even if based on C, like C++ ([Str86]), we can find a much greater concern about the number of arguments to a function and the type of identifiers (for example, the declaration of a function must necessarily contain the type of the arguments, so our compar formal C argument is not a valid formal C++ argument), that leads to a stronger type system. Unfortunately, unsafe pointers are still there (it is possible to rework the C example above to compile correctly in C++).[6]

1.3.2 Limits of static type-checking

Even if there are many programming languages that are able to provide safe static type-checking systems, it is important to remember that there are important programming constructs for which we do not know how to provide a safe static type system. A typical example is the array data type: if a program tries to access the 35th element of an integer array of only 20 elements, the result is an error that can very well be considered as a type error, since the index of an array in this case should morally belong to the type of the integers in the range 1 to 20, and 35 is of a "different" type. Nonetheless, the index of an array is usually computed by the program itself, and the problem of knowing if this index is inside the array boundaries is in general undecidable. The only viable solution then is to check for the correctness of the array index at run-time.[7] A very similar and common situation occurs with operations that are not defined on all the values of a given type, like the division, that raises an exception if its second argument is 0.

---

[5]Indeed, there is a separate program lint that tries to signal at least the most blatant atrocities in C programs, but this is more a pointer to the problem than a solution to it.

[6]We simply declare the type of the arguments of the formal parameter compar to be char *, and then pass as actual parameter a function taking as arguments pointers to whatever data structure we want.

[7]A different, non-automatic solution is provided by an external proof that the index will always be in the boundaries.

1.3.3   Functional languages

While imperative languages made their evolution, declarative languages
(first just functional ones, and later on also logical ones) gained broader
acceptance. In declarative languages we can find a major effort to depart
from the original Von Neumann computational model: notions like *state*,
memory and the like disappear or are given limited citizenship, while one
focus on a higher-level description of *what* a program is supposed to do,
instead of the precise details of *how* to physically handle some pieces of
data. Declarative programs are easier to write, read, and show correct than
their imperative analogues, but this came at the beginning at the price of
a much worse time and space efficiency, so that they were originally used
only for artificial intelligence or rapid prototyping and were spread out
much less than the imperative languages. Nonetheless, the greater purity
of functional languages allowed to develop for them a variety of highly
powerful and flexible type systems, which are the current state-of-the-art
in this domain. As far as efficiency is concerned, let us notice here in passing
that things are really changing now: see for example [DVR92, Mac93] for
recent advances.

In the functional programming community, on one side we can find a
wide community of users that adopted Scheme: this is a modern version of
the original Lisp language, that, even if still almost untyped, has been influ-
enced by the growing need of automatic consistency checks, and implements
a limited form of dynamic type-checking.[8] Here is how the declaration of
a sort function would look like in Scheme:

```
(define (qsort cmp l )
  . . .
)
```

On the other side, we find another community of users that has com-
mitted to a group of strongly typed functional languages all based on the
original ML language introduced by Milner in the 70s as a meta language for the LCF proof system. Hope [BMS80, FH88], ML [MTH90],
Miranda [Tur85], Haskell [HPJe92], LML [AJ89], Standard ML [MT91,
MTH90], CAML [CH88, WAL+90] and CamlLight [LW93, L+93, WL93]
are all based on the same idea: The language is strongly typed, with every
expression getting a proper type that the system can infer for the programmer (he is not obliged to give any annoying type annotation) and a program
for which the system can infer a type will never produce a type error at run
time.

---

[8]In Common Lisp, there is an extensive machinery available to the user to *declare*
types, but it is entirely optional.

Here is how our quicksort routine looks like in CamlLight, a highly portable implementation of a dialect of ML (something very similar to this code can be written in Scheme):[9]

```
#let rec qsort cmp =
#    let rec split a =
#        fun   [] -> [],[]
#        | (b::r) -> let (m,M) = split a r in
#                      if (cmp (a,b)) then (m,b::M)
#                                     else (b::m,M)
#    in fun   [] -> []
#       | (a::l) -> let m,M = split a l
#                   in (qsort cmp m)@(a::(qsort cmp M))
#;;
qsort : ('a * 'a -> bool) -> 'a list -> 'a list = <fun>
```

To sort a list a::l of $1 + n$ elements[10], the routine qsort first uses the local function split to obtain two smaller lists m and M of elements respectively lesser and greater than a, then recursively sorts them, and finally puts a back in place between (qsort cmp m) and (qsort cmp M). This program corresponds directly to the high-level description of a quicksort algorithm that one can find in a usual textbook: the proof of correctness of the program is as easy as that of the algorithm.

Notice that even if the user is not required to fill in any type declaration at all, the type system is able to infer for the program a very general type: the 'a in the *inferred* type is a *generic* type variable, which can be instantiated at any other type. This tells the user that his quicksort routine (actually, its very code) can be used with lists of any given type.

1.3.4   The lambda calculus

What is surprising in functional languages is that they can be seen as syntactic sugared versions of an extremely simple mathematical formalism: the λ-*calculus*. In this formalism, everything is an (untyped) function, and the syntax is extremely sober:

$$f := x \mid ff \mid \lambda x.f$$

A function can be a variable or an application of a function to another one, or we can build a new function from an old one by *abstracting* a given

---

[9]The # identifies user input, and the last line is the system's output, in this case, the type of qsort.

[10]In ML, :: is the operation of adding one element at the beginning of a list, while @ is the usual list concatenation.

variable x in it using the λ operator (this corresponds to the declaration of a formal argument in an imperative language). There is only one very simple operation allowed, and that is the *substitution* of an actual argument for a formal argument.[11]

$$(\beta) \qquad (\lambda x.f)g \longrightarrow f[g/x]$$

This says that whenever a function declaring a formal argument x is passed an actual argument g, we just *substitute* x with g wherever x appears in f.

It was originally designed by Church as an alternative foundation of mathematics [Chu32], based directly on *functions* instead of *sets* as in Frege's work, but very soon Kleene and Rosser [KR35] came up with a paradox that showed that the λ-calculus is logically inconsistent. While we refer the interested reader to [HS80], Ch. 17, for a nice presentation, let's just recall that a very simple version of the paradox uses a particular function Y that is able to compute the fixpoint of any other function f, i.e., such that $Yf = f(Yf)$. Using Y it is easy to prove *any* proposition in Church's logical system based on λ-calculus that is then inconsistent.

In parallel with what happened to Frege's formalism, also for the λ-calculus there have been successful tentatives to rule out the paradoxes by adding types to it. In the simply typed λ-calculus, for example, Y is not a legal function.

But if we drop logic and look only at computations, this very paradoxical function Y has enormous value: it allows us to build recursive programs without any special syntax.

For example, take a simple recursive definition computing the factorial of a given integer, and perform the following formal manipulation:

```
let rec fact =
 (function n -> if (n = 0) then 1 else n*(fact (n-1)));;

let rec fact =
 (function f ->
   function n -> if (n = 0) then 1 else n*(f (n-1))) fact;;
```

Then we see that `fact` is the fixpoint of the function

```
(function f ->
  function n -> if (n = 0) then 1 else n*(f (n-1)))
```

and we can simply declare that `fact` *is* the function

---

[11]For a more formal presentation, see [Bar84].

```
Y (function f ->
     function n -> if (n = 0) then 1 else n*(f (n-1)))
```

So, we can write recursive functions directly in λ-calculus without any real difficulty.

Indeed, it turned out very soon that the λ-calculus has the same computational power as Turing machines [Tur37], and any other universal programming language, while retaining an extreme simplicity [Rog88].

Also, even if it is not a full-blown programming language itself, the λ-calculus contains many fundamental problems of function call and application in a pure form. For this reason, along the years, it rapidly became a fundamental tool to study programming language in general and functional programming languages in particular.

1.4 Exploring typed λ-calculi

Type systems for programming languages can very well be studied in the λ-calculus, seen as a core functional language. For example, the untyped λ-calculus that we just met corresponds roughly to Lisp, with functions that can be applied and passed as arguments without any restrictions, but we can modify it here and there to get type systems similar to those of Pascal, ADA, or even much better than those.

1.4.1 Church-style types

We can easily get a strongly typed monomorphic language by decorating terms with types and imposing restrictions on the function application, as follows:

$$\Gamma, x : A \vdash x : A \quad (VAR)$$

$$\frac{\Gamma, x : A \vdash g : B}{\Gamma, \vdash (\lambda x : A.g) : A \to B} \ (ABS) \qquad \frac{\Gamma \vdash f : A \to B \qquad \Gamma \vdash g : A}{\Gamma \vdash fg : B} \ (APPL)$$

Here what is to the left of the *entailment* symbol $\vdash$ is an *environment*, that is, a sequence of type declarations for the variables, with Γ ranging over environments. The rules allow us to give a unique type to any legal program and to rule out programs that cannot be typed. The first rule tells us that a variable x declared of type A really has that type; the second rule allows us to define new functions by declaring a new formal parameter x *and* its type; the third rule just says that the type of the actual argument of a function f must match the type of the formal argument of f.

This type system (known as typed λ-calculus *à la Church*) is strong, explicit and monomorphic: we are forced to declare the type of the arguments of any function, and this type will stay the same along all the computation. As in Pascal, we are forced here to rewrite the very same function as soon as the type of the argument changes. For example, even for the identity functions, which just passes back its argument without even touching it, we must specify the type of the argument as in $Id_{int} = \lambda x : int.x : int \to int$, and then we cannot apply Id_{int} to characters, even if it would work flawlessly on characters also. It is very annoying: we cannot even cut and paste our code, since the type-checker would not let it go without properly changing the type declarations for the formal parameters!

1.4.2   Curry-style types

We get a more flexible system by making it implicit. It is enough to change just the abstraction rule, taking out the type declaration for the argument

$$\frac{\Gamma, x : A \vdash g : B}{\Gamma, \vdash (\lambda x.g) : A \to B} \quad (ABS_{impl})$$

The resulting type system (known as typed λ-calculus *à la Curry*) is still strong and allows us to write something like $Id = \lambda x.x$, which can be applied, without fiddling with its definition, to an argument of any type. But the task of verifying if a program is correct is now harder: when we find a function definition, we have to *guess* the right type of the formal argument from the context. For example, if our program is Id 3, then 3 is of type *int* and we can deduce that the x in the code of Id must be considered of type *int*. Things can get complex, but it is possible to do this guessing in an efficient way. Still, this system is essentially monomorphic: we can reuse code at different types by cut and paste without modifying it, but we cannot use at different types a function passed as argument. Take $(\lambda xy.y(x3)(x"a"))Id$: it will not type-check, because once we make our choice for the type of Id, then the formal parameter x also gets that type, and it cannot be both $int \to int$ to be compatible with 3, and $char \to char$ to be compatible with $"a"$.

1.4.3   Explicit polymorphic types

To get a polymorphic system, we must use more powerful types, rather unfamiliar to the usual programmers, that allow to do something similar, but much more powerful than the `generic` construct of ADA. The well-known polymorphic λ-calculus (or System F) [Gir72, Rey74] is obtained by adding to the monomorphic system just the following two rules to generalize and instantiate types:

$$\frac{\Gamma \vdash M : A}{\Gamma \vdash \Lambda X.M : \forall X.A} \ (GEN) \ ^{12} \qquad \frac{\Gamma \vdash M : \forall X.A}{\Gamma \vdash M[B] : A[B/X]} \ (INST) \ ^{13}$$

This type system is strong, explicit and polymorphic: the new syntax for types and terms allows easily now to reuse the same code at different types, even passed as argument to other functions. We can write our identity function as $Id_{poly} = \Lambda X.\lambda x : X.x$, that gets type $\forall X.X \to X$. This says that Id_{poly} can be used with type $X \to X$ for *any* possible type X.

Let's come back to our previous simple example $(\lambda xy.y(x3)(x"a"))Id$: using polymorphic types, we can easily rewrite it as

$$(\lambda x : (\forall X.X \to X).\lambda y : (int \to char \to A).y(x[int]3)(x[char]"a"))Id_{poly}$$

and then it is easy to check that it is a legal program: the (GEN) and (INST) rules allow us to first declare that a piece of program can be used at different types and then actually use it at different types. Checking type correctness is still quite easy, because the programmer must annotate every formal parameter with its type and explicitly declare when a piece of code is polymorphic and how to instantiate it when he wants to use it.

Notice how the abstraction over types $(\Lambda X....)$ is analogous to the **generic type** $X...$ declaration we saw in ADA and the $x[int]$ construct is similar to **new x(int)** in ADA, but there is a fundamental difference. In ADA, it is not possible to pass around a **generic** function as an argument while here it is quite natural, and is what we just did in our example with Id_{poly}; also, in ADA any **new** instruction makes a copy of the original code, while here it is really the same code that is executed. The polymorphic λ-calculus has indeed a type system much more flexible and powerful than most actual programming languages to date.

1.4.4 Implicit polymorphic types

Then, why not get the best of all by taking an implicit version of System F? We would like to get the power of this polymorphic type system, but it is a hassle to write down types all the time. So what about having the compiler fill in the types for us? The problem is that this filling-in-the-details is easy for monomorphic systems but turns out to be impossible

---

[12]This introduction is intended to be a simple presentation, that avoids technicalities, but let's remark that (GEN) is legal only if X is not free in the type of any free variable of the term M. A formal presentation is given in 1.5.1.

[13]for any type B

for System F [Wel]. Indeed, to design actual programming languages with
similar features (flexible typing and type-inference), it turned out that it
was possible to use a stripped-down version of System F's polymorphism,
where the implicit version of (GEN) and (INST) rules are allowed only in
specific places: This is done by formulating the typing rules using two kind
of types: *monotypes* (usually noted τ), which do not contain any quantifier,
and *type schemas* (usually noted σ), which can only have a prefix of type
quantifications.

(VAR) $\qquad \Gamma, x : \sigma \vdash x : \sigma$

(INST$_{impl}$) $\quad \dfrac{\Gamma \vdash M : \forall X.A}{\Gamma \vdash M : A[B/X]}$

(GEN$_{impl}$) $\quad \dfrac{\Gamma \vdash M : A}{\Gamma \vdash M : \forall X.A} \quad X$ not free in Γ

(ABS$_{impl}$) $\quad \dfrac{\Gamma, x : \tau_1 \vdash g : \tau_2}{\Gamma, \vdash (\lambda \mathbf{x}.\mathbf{g}) : \tau_1 \to \tau_2}$

(APPL) $\quad \dfrac{\Gamma \vdash f : \tau_1 \to \tau_2 \qquad \Gamma \vdash g : \tau_1}{\Gamma \vdash \mathbf{fg} : \tau_2}$

(LET) $\quad \dfrac{\Gamma, x : \sigma \vdash f : \tau \qquad \Gamma \vdash g : \sigma}{\Gamma \vdash \mathtt{let\ x=g\ in\ f} : \tau}$

Table 1.1: The typing rules for ML (Damas-Milner style)

With this restriction, which essentially allows to use the polymorphism
of (GEN) and (INST) only in the (LET) rule, type-inference becomes
decidable.[14]

This system, due to Milner [Mil78], was originally used to program the
LCF theorem prover, but it is now the core of a large and rapidly evolving
family of modern functional programming languages. It provides up to now
the best of both worlds: the safety of typeful programs and the friendliness

---

[14]And also quite efficiently in practice, even if it takes exponential time in the worst case[Mai90].

of a syntax where we are not forced to write types. In more recent works on ML (like [MTH90, Ler92]), it is more often used an alternative (even if equivalent) presentation of these basic typing rules that is *syntax-directed*[15] and hence easier to use in the proofs. This alternative presentation will be introduced and used later on in Chapter 6.

⚠ 1.5 The typed λ-calculi used in this work

It is now time to introduce the explicitly typed calculi that we will use in the following chapters, while deferring the treatment of implicitly typed calculi to Chapter 6. The formalism chosen to present them is the Natural Deduction style, which has the great advantage to make self-evident the connection between typed λ-calculus and proofs in Intuitionistic Logic known as Curry-Howard Isomorphism and presented in the next section.

We give the full definition of the calculus $\lambda^2\beta\eta\pi*$, the most complex of the four we consider, while for the subsystems of it we will just detail the differences.

1.5.1 The calculus $\lambda^2\beta\eta\pi*$

- Types are defined by the following grammar:

  $Type ::= At \mid Var \mid Type \rightarrow Type \mid Type \times Type \mid \forall X.Type$

  where At are countably many atomic types including a distinguished constant type **T** and Var countably many type variables. Usually $\rightarrow$ is referred to as the *arrow* type, and $\times$ as the *product* type. The intended meaning of **T** is the terminal object in the categorical sense, so * below will stand for the unique term of type **T** (as required of a terminal object).[16]

- Terms ($M : A$ will stand for M *is a term of type A*)

  - the set of terms contains a countable set $x, y, \ldots$ of term variables for each type and a constant *:**T**

---

[15]This term is used to design those typing systems where it is enough to look at the syntax of a term to determine which rule must be used. This is not the case here, as the use of (GEN) and (INST) does not depend on the syntax of the term to type.

[16]This notation is different from the one originally used in [BDCL92], where the symbol $*_A$ stands for the unique arrow of type $A \rightarrow$ **T**, and, though completely equivalent and interchangeable, is preferred here for ease of reference to [CDC91], where confluence of related systems is studied.

– terms are constructed from variables and constants via the following lambda abstraction, application, pairing, projection, universal abstraction and universal application rules (notice the perfect analogy with the introduction and elimination rules for second-order logic in natural deduction style)

$$\frac{\Gamma, x : A \vdash M : B}{\Gamma \vdash \lambda x.M : A \to B} \qquad \frac{\Gamma \vdash M : A \to B \quad \Gamma \vdash N : A}{\Gamma \vdash (MN) : B}$$

$$\frac{\Gamma \vdash M : A \quad N : B}{\Gamma \vdash \langle M, N \rangle : A \times B} \qquad \frac{\Gamma \vdash M : A \times B}{\Gamma \vdash \mathrm{p}_1 M : A} \qquad \frac{\Gamma \vdash M : A \times B}{\Gamma \vdash \mathrm{p}_2 M : B}$$

$$\frac{\Gamma \vdash M : A}{\Gamma \vdash \lambda X.M : \forall X.A} \,^{17} \qquad \frac{\Gamma \vdash M : \forall X.A}{\Gamma \vdash M[B] : A[B/X]} \text{ for any type } B.$$

Notice that pairing and projections are new *term formation rules* and not constants added to the language.

• Equality

$$(\beta) \quad (\lambda x.M)N = M[N/x] \quad (\eta) \quad \lambda x.Mx = M \text{ if } x \notin FV(M)$$

$$(\pi) \quad \mathrm{p}_i \langle M_1, M_2 \rangle = M_i \quad (\mathbf{SP}) \quad \langle \mathrm{p}_1 M, \mathrm{p}_2 M \rangle = M$$

$$(\mathbf{top}) \quad M = * \text{ if } M : \mathbf{T}$$

$$(\beta^2) \quad (\lambda X.M)[A] = M[A/X] \quad (\eta^2) \quad \lambda X.M[X] = M^{18}$$

We will write $=_{\beta^2 \eta^2 \pi *}$ for the equality generated by β, η, π, SP, top, β^2 and η^2.

This presentation of our calculi differs slightly from the one used in presenting the untyped λ-calculus: instead of a *reduction* rule, we give *equalities* between terms. Let us remark that, as we will discuss at greater length in the following chapter, the two presentations are basically interchangeable, but using equalities gives a more abstract and general view.

1.5.2   General notations for terms and substitutions

Here is a good place to recall some very common notation used when dealing with λ-calculus. These will come in handy in the more technical sections of this book.

---

[17]With the proviso that the type variable X is not free in the type of any free variable of the term M.

[18]With the proviso that X is not free in M.

Notation 1.5.1 (sequences, substitutions, variables)

We will often use sequences of variables $[x_1,\cdots,x_n]$ or terms $[M_1,\cdots,M_n]$, noted respectively $\vec{x}$ and $\vec{M}$.

The *length* of a sequence $\vec{M}$ of terms (or variables) is the number of elements in the sequence, noted $\mid \vec{M} \mid$. A sequence can be empty. As is standard notation in the theory of λ-calculus, we will often use $\lambda \vec{x}.\vec{N}$ as a shorthand for $\lambda x_1 \cdots x_n.(\cdots(N_1 N_2)\cdots N_m)$, where it is intended that N_1 is not an application $N_{11} N_{12}$, as otherwise we could take N_{11} as the starting term of the sequence.

Given a term N, a sequence $\vec{M} = [M_1,\cdots,M_n]$ of terms and a sequence $\vec{x} = [x_1,\cdots,x_n]$ of variables, of the same length, $N[\vec{M}/\vec{x}]$ denotes the simultaneous substitution of every variable x_i with the term M_i in the term N. For simplicity, we always assume that bound variables are renamed as necessary to avoid capture of free variables. We may also use $\vec{N}[M/\vec{x}]$ for the simultaneous substitution of all the variables in $\vec{x}$ with the same term M (similarly for type variables).

Let $I_A = \lambda x : A.x$ be the identity of type A. As usual, by $FV(M)$ we mean the free term variables $x : A$ of a term *declared with their type*, but, being in a second-order setting, we also use $FTV(M)$ for the free type variables of a term M. Since the calculus is explicitly typed, we can write $Type(M)$ for the type of the term M, which is uniquely determined once we are given the type of its free variables.

Our calculi also have pairs, so we use abbreviations for nested pairs, or n-tuples.

Notation 1.5.2 (n-tuples)

We write $\langle M_1, \cdots M_n \rangle$ for $\langle M_1, \langle M_2, \langle \cdots, \langle M_{n-1}, M_n \rangle \rangle \cdots \rangle \rangle$.

Subsystems of $\lambda^2 \beta \eta \pi *$

The other calculi we are interested in can be naturally defined as restrictions of $\lambda^2 \beta \eta \pi *$: to obtain them we reduce the class of types and/or terms, and accordingly redefine the equality.

- The calculus $\lambda^2 \beta \eta \pi$ is $\lambda^2 \beta \eta \pi *$ without unit types. (Equality for $\lambda^2 \beta \eta \pi$ will be noted $=_{\beta^2 \eta^2 \pi}$ and is generated by β, η, β^2, η^2, π and SP.)

- The calculus $\lambda^2\beta\eta*$ is $\lambda^2\beta\eta\pi*$ without product types, pairing and projections. (Equality for $\lambda^2\beta\eta*$ will be noted $=_{\beta^2\eta^2*}$ and is generated by β, η, β^2, η^2 and *top*.)

- The calculus $\lambda^1\beta\eta\pi*$ is $\lambda^2\beta\eta\pi*$ restricted to the first order or simple types. (Equality for $\lambda^1\beta\eta\pi*$ will be noted $=_{\beta\eta\pi*}$ and is generated by β, η, π, SP and *top*.)

- The calculus $\lambda^1\beta\eta\pi$ is the restriction of $\lambda^1\beta\eta\pi*$ obtained by removing the unit type, pairing and projections. (Equality for $\lambda^1\beta\eta\pi$ will be noted $=_{\beta\eta\pi}$ and is generated by β, η, π and SP.)

- The calculus $\lambda^1\beta\eta*$ is the restriction of $\lambda^1\beta\eta\pi*$ obtained by removing product types, pairing and projections. (Equality for $\lambda^1\beta\eta*$ will be noted $=_{\beta\eta*}$ and is generated by β, η and *top*.)

1.6   The Curry-Howard Isomorphism

We have just seen that typed λ-calculus is very useful to compare in a uniform framework different type systems for programming languages: it provides a simple, yet powerful foundation. But this is not the only nice surprise. Consider the following λ-terms:

$$K \;=\; \lambda x.\lambda y.x$$
$$S \;=\; \lambda x.\lambda y.\lambda z.(xz)(yz)$$

These terms are very well known to λ-calculus experts. Without entering into too much detail, let's just say that they contain in a sense the essence of functionality and provide the basis of Combinatory Logic (see [HS80] for a nice presentation of the topic).

Let us have a look at their types. It is possible to compute them by hand, but nowadays you can use your preferred ML system to do this (we use CamlLight [LW93, L$^+$93, WL93]):

```
#let K = fun x -> fun y -> x;;
K : 'a -> 'b -> 'a = <fun>

#let S = fun x -> fun y -> fun z -> (x z) (y z);;
S : ('a -> 'b -> 'c) -> ('a -> 'b) -> 'a -> 'c = <fun>
```

In the late 60s Curry and Howard noticed a very intriguing analogy between the types of these important λ-terms and the axioms of intuitionistic positive propositional logic. Indeed, if we consider the arrow not as a function space, but as a logical implication, and then look at these types as if they were logical formulae, they turn out to be true propositions, and very fundamental ones: using the types of S and K as propositions we can derive all true formulae of intuitionistic positive propositional logic (these types are *the* proper axioms of this logic [BM77, How80]).

This initial observation lead to the discovery that there is a strong correspondence between typed λ-calculus and intuitionistic logical systems, that is very well summarized in a slogan that has been around for a long time in the lambda calculus community: *types are propositions and programs are proofs*. To any proposition we can associate a type by simply changing implication symbols into arrows, and to any logical proof of this proposition we can associate a λ-term that has that type. And vice-versa. This analogy carries on to computations: the basic computational mechanism of λ-calculus, the β-reduction rule we have seen above, exactly corresponds to the traditional cut-elimination procedure used to normalize proofs (see for example [GLT90]).

This surprising correspondence between λ-calculi and logical systems was noticed by Curry and Howard, and it is now universally known as the *Curry-Howard isomorphism* or the *formulae as types* paradigm. Their notes had been widely circulated long before being finally published in [How80], and they had enormous influence on the successive development of the theory of λ-calculus and functional programming.

The Curry-Howard isomorphism provides a nice basis with which to deal with the issue of *program correctness*, by relating data types manipulated by programs to logical propositions used to prove program correctness. In this area it is possible to identify many different approaches, and for each of them there is a great deal of work already done in the literature. Still, I think it is rather fair to divide them into approaches to *total* and *partial* correctness. We will briefly recall here some of the more paradigmatic ones to provide a rather general framework where to fit the material presented in this book, but without any pretension of completeness.

Total correctness

The idea is that we would like to guarantee not only that our program will not attempt to do something foolish, but also that it does exactly what we *expect* it to do. To formalize our expectation, we usually resort to some kind of a specification language that is essentially algebraic or logical, and we match the specification against the program, either *before* or *after* actually writing the program, thus taking one approach somewhere in between the following two extreme ones.

- **Program Verification**
  One can verify that a program is not going to exhibit unwanted behav-

ior *after* writing it by means of a formal logical language suitable to
verify certain soundness conditions, typically the preservation of some
invariants. This seems by now a well-established area in the field, and
such techniques are actually used in program verification (see [Gri83],
[AEF89]). The major defect I see in this approach, though, is that the
programming language itself and the logic meta-language used to perform the verification are often two completely unrelated formalisms:
as a consequence, the task of verification demands the knowledge of
two languages and requires a fair effort from the human being who
has to guess the "right" invariants.

• **Extraction of programs from specifications**
One can completely reverse the problem and derive a program starting directly from a specification. Relying on the Formulae as Types
paradigm, the program will be automatically extracted from a proof
of the specification seen as a formula in a suitable constructive logic.
This approach has the great advantage that the correctness of a program is guaranteed by the very way it is extracted, and the programmer needs to know only the logical language, that in any case is
closely related to the programming language (it is just its algorithmic
counterpart). But there is currently a major drawback: there is a
serious lack of control on the algorithmic behavior of the extracted
program, and before envisaging any practical application, the issue of
efficiency must be addressed satisfactorily. In any case, there is a serious effort in this direction, that seems quite promising (see [KP90],
[Kri90], [C+86b], [How88], [PM89], [CH85], [Bas89]).

Partial correctness

Since it has unfortunately well known for a long time that we cannot design
a tool to verify that an arbitrary program satisfies an arbitrary specification, both of the previous approaches must trade in something in order to
prove total correctness of a program. The first one relies on a human being
(playing the role of an *oracle*) to find the right invariants to check, while the
second, besides relying on a human being to find the proof from which to ex-
tract the program, typically also restricts the class of computable functions
that are programmable[19] and the class of algorithms for the programmable
functions.[20]

---

[19]In the Calculus of Constructions, for example, only the functions provably total in
the arithmetic of any order can be coded. This is a wildly big class of functions, but it
is well far from containing all the Turing-computable functions.

[20]For any function, there are many possible different algorithms, some more efficient
than others: in System F it is not possible to compute $n-1$ in constant time [Col91].

Checking total correctness can become so cumbersome that we easily feel compelled to resort to some kind of a less demanding but more manageable notion of correctness, where a little part of our freedom in writing programs is traded in just for a guarantee that a legal program will not allow generally unwanted behaviors like the ones we previously discussed. We want some kind of automatic system that performs the verification without external help: his task must be a decidable one. So we leave program verification/extraction for type-checking (or type-inference). In a sense, we ask to our language to be more *intelligent*, but not too much. It must be able to check certain soundness conditions and to allow only programs that are guaranteed to satisfy them, but we do not require from it a guarantee that the computed function is exactly what we had in mind. Borrowing from physics, we can say that we will be satisfied to know that the dimensions of an equation are okay, even if this does not prove that equality holds.

We can take up again the Formulae as Types paradigm, and we look for a restriction of the notion of correctness to be obtained by reducing the class of formulae legal in the logic. The natural choice is to forbid first-order formulae in our logic and work only with the higher-order fragment. This way, we can no longer fully describe the algorithmic content of a program, but just its skeleton, i.e., the higher-order logical formula derived from the full specification by erasing all the first-order components, so that we avoid the general undecidability results. We will not be able to perform program extraction, so the program must be written by the user, but our formal system will be able to check it against the (now *partial*) specification.

Let's see this on a simple example. Imagine that we specify a function add that performs the addition of two integer arguments:

$$\text{add} : \forall x. \forall y. \text{int}(x) \Rightarrow \text{int}(y) \Rightarrow (\text{int}(add\ x\ y) \land (add\ x\ y) = x + y)$$

To verify that an actual implementation of add satisfies the specification, we have to prove a theorem stating that whenever x and y are integers, then so is add $x\ y$ and also (add $x\ y$) $= x + y$: a difficult task. But if we drop all the first-order part (that is in italic in the specification above), we are left with the following partial specification (that will sound more familiar):

$$\text{add} : \text{int} \Rightarrow \text{int} \Rightarrow \text{int}$$

If we look at this formula as a type, it is indeed the type of an integer-valued functions that takes two integer arguments. This partial specification can be provided by the user via type declarations *as part of the program* (in such case we say that our system performs *type-checking*), or our system can try to deduce the right specification (better if *all* partial specifications satisfied by the given program) from the way the program is used (and then we get a *type-inference* system, like the ML one).

Indeed, starting from a different, logical point of view, we are led again to the ML language as a reasonable compromise between the burden imposed on the programmer by the type system and the degree of correctness of the program that can be verified by the system.

1.7   Using types to classify and retrieve software components

The problem of easily retrieving existing software components from a library is a fundamental one. If it is easier to rewrite a function than to find an existing one that does what we need, then that piece of software will be rewritten by many programmers, wasting time, and probably will be rewritten differently, endangering any standardization effort. The construction of large libraries of simple functions will not lead, then, to that code reuse that is clearly seen as a fundamental tool to master very large software projects. This is especially true in the functional programming and in the object-oriented communities, where the code associated with a single function or class is usually small (and easy to rewrite), while the number of available functions and classes is very large.[21]

The need for a fast, simple and intuitive way to retrieve existing software components has been growing up also for programming languages that are more widely used than strongly typed functional languages, but the long tradition of paper documentation - along with the huge amount of books, tutorials and guides available for the standard libraries of these languages - have made difficult to distinguish between two very different classes of information:

- what the programmer should *learn* by himself about a system, and

- what the system should be able to *retrieve* for the programmer.

In fact, today it is rather commonplace to work with systems where the programmer has to learn everything (maybe by reading some multi-hundred page tutorial to programming with C++), and the system does not provide even the simplest tool to search for a piece of code. In this author's opinion, a programmer should *learn* only the few high-level concepts of a

---

[21]The standard library of CAML v.3.0 contains already more than 1000 user-level identifiers, and the Application Kit of NEXTSTEP (the smaller and best engineered object-oriented library known to this author) contains more than 100 classes and 2000 methods.

programming language, like scoping rules, control mechanisms, data model, abstraction facilities, inheritance mechanisms and so on: this is a task that humans are good at. He should definitely not learn, but *retrieve*, with the help of the system, all information about the thousands of useful library routines that are already available to read input, write output, perform complex data manipulation and the like: this is a task that machines are good at.

High-level concepts about a given programming language evolve slowly over the years, while huge new libraries can be created in a few months. The former must be taught, but for the latter the only viable solution is to build efficient retrieval tools. As has been said by many people, retrieving existing software should be easier than writing it from scratch.

1.7.1 Object-oriented languages

This situation is especially evident for those languages used to build large object-oriented libraries, where the ease of writing many very simple classes of objects instead of a few huge monolithic procedures can rapidly lead to loose control of the existing software base.

There has been some work in the past years on the use of the *class types* information available in object-oriented systems to *classify* the elements of the system libraries. In these systems, the notion of *inheritance* and that of *subtyping* of classes (that are different notions, as first shown in [CCH90]) provide a simple starting point to present organically the information contained in an object library. Inheritance is used in the NEXTSTEP system as a basis for the `HeaderViewer` browser tool available in the NEXTSTEP developer package and shown at work in Figure 1.1.

Subtyping, or *conformance*, is used instead for the analysis of a part of the Smalltalk-80 ([GR83]) library in [Coo92], together with specification techniques developed earlier for objects by America in [Ame91]. This study is very promising, as it also points out how the implementor's and the user's views of a system can be very different. It would be possible to build a search tool similar to `HeaderViewer` based on this notion.

Despite these recent efforts, though, the tools generally available today are still nothing more than a prehistoric alphabetical index of identifiers, maybe with some facility to enable regular-expression searches (like in the CAML interpreter, see [CH88]) or some kind of thesaurus, useful when you have to find your way in an Unix manual (the well-known -*k* option of the *man* command).

But the name given to a function (or a method) is left to the more or less original imagination of the programmer, so if you change systems, you change dialect also. Borrowing an amusing example from [Rit90b],

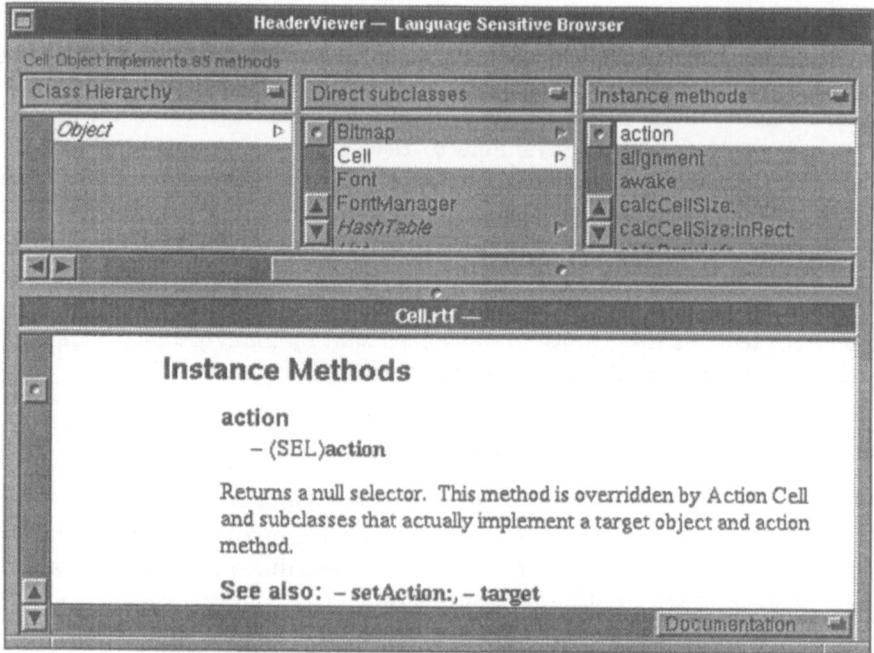

Figure 1.1: NEXTSTEP `HeaderViewer` Class Browser.

if we look for a function that distributes a binary operation on a list,[22] we can easily collect in the different functional programming languages a nice amount of names: *itlist, list_it, foldl, fold* and *fold_left*, so that the rudimentary tools available to search the libraries do not help at all.

For this reason, also the current subtyping and inheritance approach to class browsing and searching still seem inadequate, as they do not take into consideration the usually arbitrary choice of names and of parameter passing conventions that we will successfully deal with for purely functional languages.

1.7.2   Functional languages

Since function *names* are quite arbitrary, we need to look for something better, and here is where *types* come to our rescue. For pure functional languages, as we will see, types provide a powerful *specification* language that can be used for searching with high precision into program libraries,

---

[22]Here is, for example, a very simple code for `list_it` in CamlLight: `let list_it f = fun [] b -> b | (a::l) b -> f a (list_it f l b);;`

while for imperative languages and object-oriented languages, there is still work to be done, even if the same technique represents already a great improvement over what is available now.

The fact that types can be seen as partial specifications of functional programs has been exploited at length in the family of ML languages to perform a partial correctness verification, but it is only very recently that another very appealing application of this analogy has been pointed out. If types are partial specifications of the behavior of programs, then they can be successfully used as keys to *search for existing programs* with a given behavior. Using the *type* of a function as a search key provides the programmer with a uniform and sensible tool to retrieve data in libraries. The *types*, with their logical counterpart, would provide the necessary standard language. This simple but rather new idea was proposed originally by Mikael Rittri ([Rit91]), Colin Runciman and Ian Toyn ([RT91]).

Language	Name	Type
ML (LCF)	itlist	$\forall X.\forall Y.(X \to Y \to Y) \to List(X) \to Y \to Y$
CAML	list_it	"
Haskell	foldl	$\forall X.\forall Y.(X \to Y \to X) \to X \to List(Y) \to X$
SML of New Jersey	fold	$\forall X.\forall Y.(X \times Y \to Y) \to List(X) \to Y \to Y$
Edinburgh SML	fold_left	$\forall X.\forall Y.(X \times Y \to Y) \to Y \to List(X) \to Y$

Table 1.2: Example of syntactically different but isomorphic types in function libraries.

They immediately notice, though, that functions that we want to consider essentially the same turn out to be assigned types that at first sight seem pretty different. For example, look at the type that the *itlist - list_it - foldl - fold - fold_left* function is assigned in five different widely used languages based on the same polymorphic type discipline originally presented in Milner's ML [Mil78] (see Table 1.2, which is borrowed from [Rit91]).

1.8 When are two types equal?

This simple example shows that the syntactic equality of types is too much a fine relation on types to be used for our purposes. So what is the *right* way to compare types? We need a coarser relation on types that allows us to consider *equivalent*, for example, all these (partial) specifications, as the associated function is the same up to some kind of reorganization of the arguments.

More precisely, for any two types A and B of the five in Table 1.2 above, we can write two simple transformations $h : A \to B$ and $h^{-1} : B \to A$ such that

- for any function $f : A$, $h(f) : B$ and $h^{-1}(h(f)) = f$, and

- for any function $g : B$, $h^{-1}(g) : A$ and $h(h^{-1}(g)) = g$.

The functions h and h^{-1} are exactly the little piece of code that a programmer writes when he finds a useful function f of type A in a library, and he wants to use it in his program in a place where a function of a "similar" type B is needed.

For example, if we happen to be an SML programmer that is porting his code to CamlLight, then SML library function `fold` used in our program would have to be redefined using the library function `list_it` as follows:

```
#let h f a = f ((fun n p q -> n (p,q)) a) ;;
h : ((’a -> ’b -> ’c) -> ’d) -> (’a * ’b -> ’c) -> ’d = <fun>

#let fold = h list_it;;
fold : (’a * ’b -> ’b) -> ’a list -> ’b -> ’b = <fun>
```

What we want to do is to identify the types whose elements can be effectively "coded" one into the other and vice-versa without any loss of information. Then a search *up to this equivalence of types* will find all the functions that can potentially serve our purposes. Clearly, this notion of equivalence must not depend on the particular implementation of the language, but only on its definition.

Once we choose an appropriate notion of equivalence of types that is decidable, we can build appropriate retrieval tools for a programmer like the one shown in Figure 1.2, which is described in Chapter 6.

There is nowadays a general agreement that the notion of equivalence of types which is appropriate for library searches is that of *isomorphic* types, a notion formalized and studied at great length by this author, Kim Bruce, Giuseppe Longo and Sergei Soloviev. Let's see why.

1.8.1 Isomorphic types

There have been several very interesting works on the subject, like [RT91, Rit90a, Mor91, DC92, Mat, RW91, Rit93, ZW93], especially after the seminal paper [Rit91] where Mikael Rittri pointed out that for library searches

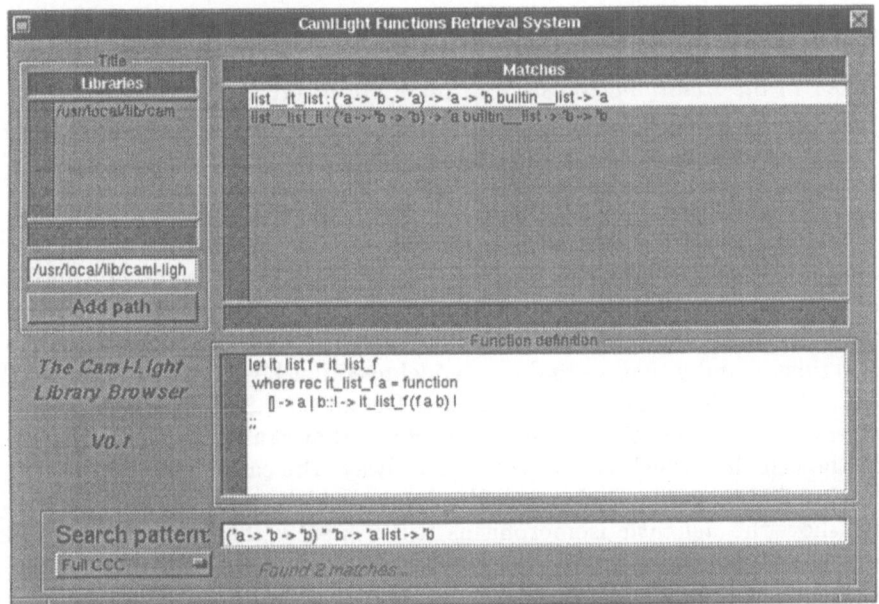

Figure 1.2: A function library browser for CamlLight based on isomorphisms of types.

it is appropriate to consider equivalent two types A and B if and only if it is possible to write for these types two conversion functions h and h^{-1} as above [Rit91]; more formally:

Definition 1.8.1 (Definable isomorphisms, invertible terms) *Two types A and B are definably isomorphic ($A \cong_d B$) iff there exist functions (λ-terms) $M : A \to B$ and $N : B \to A$ such that $M \circ N = I_B$ and $N \circ M = I_A$, where I_A and I_B are $\lambda x : A.x$ and $\lambda x : B.x$, the identities of type A and B. The terms m and N are called* invertible. *We also write $M : A \cong_d B : N$ when we want to make the associated invertible terms explicit.*

Here we start using $\circ$ as infix notation for the usual function composition combinator $\lambda x.\lambda f.\lambda g.(f(gx))$.

There is also a semantic alternative to the syntactic notion of isomorphism of types. When we provide models of typed λ-calculi, one usually interprets types as sets with structure, and terms $M : A \to B$ as structure preserving functions from the interpretation of A to that of B. So we can consider the following definition.

Definition 1.8.2 (Semantic isomorphisms) *Two types* A *and* B *are isomorphic in a specific model* $\mathcal{M}$ *if their interpretations are isomorphic in* $\mathcal{M}$ *in the traditional sense (i.e., there are in the model invertible functions* f *and* g *between them), and then we write* $\mathcal{M} \models A \cong B$. *Two types will be* semantically isomorphic, *noted* $A \cong B$, *if* $\mathcal{M} \models A \cong B$ *holds for every model* $\mathcal{M}$ *of the calculus.*

What then is the relation between the semantic notion of isomorphism of types holding in a specific model of the calculus and the syntactic notion of *definable* isomorphism? In principle, the class of the types isomorphic in every model is larger than that of definable isomorphic types, since the functions f and g that we find in a model need not be definable by terms of the calculus. The adjective *definable* in Definition 1.8.1 is really meant to stress the fact that $A\cong_d B$ is an isomorphism that can be defined *uniformly* in the calculus, which is not necessarily always the case.

In all the cases considered in this book, it turns out that one can easily show the *definable* isomorphisms to be exactly the ones that hold in every model of the calculus, and this fact establishes an important connection between the identification of types that it is desirable to make from a programmer's point of view and the notion of *isomorphisms* of objects in models of λ-calculus.

Theorem 1.8.3 *Let* A *and* B *be types. Then* $A \cong B \iff A\cong_d B$.

Proof. ($\Leftarrow$) trivial.
($\Rightarrow$) Take the open term model of the calculus. $\square$

We can then speak simply of *isomorphisms*.

The relevance of isomorphisms of types in specific models is very well known in Denotational Semantics, where one of the major successes has been the ability to build models of typed calculi that validate domain equations like $D = A + [D \rightarrow D]$, i.e., where D and $A + [D \rightarrow D]$ are isomorphic. This result is a central tool in giving semantics to programming languages, in particular to type-free languages.

While a definable isomorphism $M : A\cong_d B : N$ does hold in every possible model $\mathcal{M}$ of the calculus (the isomorphism in $\mathcal{M}$ being provided by the interpretations of M and N), an isomorphism holding in a specific model usually does not hold in *all* the models of the typed λ-calculus and is not *definable*.

There has been a great deal of investigation carried on in the last decade concerning isomorphisms of types that do hold in every model of a given typed λ-calculus. Specifically, one is interested in deciding if two given types are or are not isomorphic in every model and in finding a theory of equality Th such that $Th \vdash A = B \Longleftrightarrow A \cong B$. Of course, for the practical purpose of library searches, this theory of equality ought better be decidable (and indeed it is, as we will see in the next chapters).

The deep connection between λ-calculus, Proof Theory, Category Theory and functional programming is surely one of the reason why there has been a widespread interest in type isomorphisms in the last decade. Since the problem of finding isomorphic types can be rephrased into related problems in these other research fields, many people looked at this problem from different points of view and with different motivations. As a result, we can have today a very powerful set of tools and techniques to treat this problem.

We will try to summarize here the connections between these different research areas that are useful for understanding isomorphisms of types better.

1.8.2 Isomorphisms in category theory

Category theory is a branch of mathematics that developed recently to provide an abstract framework for stating many general results in different fields [ML71].

Definition 1.8.4 (Category) *A category C is given by*

- *a collection of objects Ob_C*

- *for any two objects A and B in Ob_C, a collection of arrows, or morphisms $C[A, B]$ from A to B,*

- *a composition operation $\circ$ s.t. for all $f \in C[A, B], g \in C[B, C]$, $g \circ f \in C[A, C]$.*

With the additional conditions that that composition is associative (i.e., $f \circ (g \circ h) = (f \circ g) \circ h)$ and that for any object A there is an associated identity arrow $id_A \in C[A, A]$ such that $f \circ id_A = f$ for all $f \in C[A, B]$ and $id_A \circ g = g$ for all $g \in C[B, A]$.

As we can see from this definition, like in λ-calculus, category theory takes the notion of function as primitive instead of that of set: we can still think of objects as sets, but we are unable to see what they "contain".

Since the only way to look into an object in category theory is to study the functions that can act on it, most of the categorical concepts end up defined up to isomorphisms, and in particular most constructions of objects. For this reason it is particularly important to be able to decide if two objects built in different ways are actually the "same" object or not, i.e., if they are isomorphic. A typical example is the case of the cartesian product, which is always commutative:

$$A \times B \cong B \times A.$$

An historical survey of results

There has been a very interesting work dedicated to the characterization of the isomorphisms holding in some special classes of categories [Sol83], even since 1981.[23] In his work, Soloviev focuses on a special class of categories, called cartesian closed categories, where for any two objects A and B there exists a *product* $A \times B$, an *exponent* $A \to B$, and a special object **T** called the *terminal* object, with a set of additional properties that make them very well suited as models of the typed λ-calculus (see for example [Sza78, LS86] and Chapter 8 of [AL91]).

If categorical tools are increasingly used to give semantics to programming languages, especially to functional ones, this is also due to a close connection between this class of categories and the typed λ-calculus. It is by now a very well known fact that the models of λ-calculus with products and unit type are exactly the cartesian closed categories (shortly, *ccc*'s), or, equivalently, that this calculus is the *internal language* of ccc's (see [Min77, LS86, AL91]).

Due to this connection, this categorical problem coincides with the characterization of isomorphisms in typed lambda calculus, which is particularly interesting for the case of ccc's. It is a simple exercise in basic category theory to show that in any ccc, the following isomorphisms hold:

In 1981, Soloviev showed, using techniques developed in number theory several years before, that these are *exactly* the isomorphisms holding in all ccc's. It is worthwhile to present here in some more details the connection with number theory used by Soloviev, as it is another interesting case of connection between very different branches of Mathematics.

If we take the table of isomorphisms of types above and we read it by taking **T** as the number 1, type variables as numerical variables, $A \times B$ as the regular product AB of two numerical expressions A and B, and finally $A \to B$ as the exponent B^A, then the isomorphisms above become very well-known numerical identities that are taught in high-school algebra classes. The reason for this is that natural numbers can be used to build a cartesian closed category: to each n we associate a finite set with n elements, and as morphisms between two objects m and n we take the n^m usual set theoretic functions between the associated finite sets.[24] In this category, the terminal object is the singleton associated with 1, while the product and exponent

---

[23]Date of the first publication (in Russian).

[24]This is usually known as the category of finite sets or finite ordinals.

$$
\left.
\begin{array}{ll}
1. & A \times B = B \times A \\[6pt]
2. & A \times (B \times C) = (A \times B) \times C \\[6pt]
3. & (A \times B) \to C = A \to (B \to C) \\[6pt]
4. & A \to (B \times C) = (A \to B) \times (A \to C) \\[6pt]
5. & A \times \mathbf{T} = A \\[6pt]
6. & A \to \mathbf{T} = \mathbf{T} \\[6pt]
7. & \mathbf{T} \to A = A
\end{array}
\right\} Th_{ccc}
$$

A, B, C can be arbitrary types and $\mathbf{T}$ is the unit type.

Table 1.3: The theories of valid isomorphisms for ccc's.

correspond directly to numerical product and exponentiation. Using these facts, it is easy to show that two objects are isomorphic *in the category of finite sets* if and only if they are equal as numerical expressions. This means that we have reduced the problem of characterizing isomorphic types *in the category of finite sets* to the problem of finding all valid numerical equation between expressions built out of product, exponentiation and the constant 1 only. This number theoretic problem is a particular case of what is known under the name of *Tarski's High School Algebra Problem*, and has as a solution exactly the set of equations of the Table 1.3 above. This remarkable fact allows us to conclude Soloviev's argument as follows: on one side, we have a set of isomorphisms that is valid in all ccc's; on the other hand, we have found a special ccc where only these isomorphisms hold. Hence, there cannot be any other isomorphism holding in all ccc's, and we are done.

⚠ 1.8.3   Digression: Tarski's High School Algebra Problem

It would be unfair here not to give at least a brief description of the works in number theory that made possible this first elegant characterization of the isomorphisms that hold in all ccc's. We will give here a general overview of the works in this field known to the author, as well as a list of references for the interested reader. In [DT69], Tarski first asked if the following usual equalities (called theory $\mathcal{E}$) that are taught in high school are complete for the natural numbers (i.e., if they are enough to prove all the true numerical equalities):

$$ab = ba \qquad\qquad (ab)c = a(bc)$$
$$c^{ab} = (c^a)^b \qquad\qquad (ab)^c = a^c b^c$$

$$a + b = b + a \qquad (a + b) + c = a + (b + c)$$
$$a(b + c) = ab + ac \qquad c^{(a+b)} = c^a c^b$$

He conjectured that they were enough,[25] but was not able to prove his conjecture. His student Martin could show that $(a^b)^c = (a^c)^b$ is complete for $\langle N, \uparrow \rangle$, and also that $ab = ba$ $\quad (ab)c = a(bc)$ $\quad c^{ab} = (c^a)^b$ $\quad (ab)^c = a^c b^c$ are complete for $\langle N, \cdot, \uparrow \rangle$, but he exhibited a simple valid equation involving sums that is not provable from these axioms:[26]

$$(x^u + x^u)^v \cdot (y^v + y^v)^u = (x^v + x^v)^u \cdot (y^u + y^u)^v$$

The question, though, was not settled by this counterexample, because it is a counterexample only if we do not allow a constant for 1, that Tarski clearly considered necessary, even if he did not explicitly mention it in his conjecture. In the presence of a constant 1, the following new equations come into play, and allow us to easily prove Martin's equality.

$$1a = a \quad a^1 = a \quad 1^a = 1$$

This problem attracted the interest of many other mathematicians, like Leon Henkin, who focused on the equalities valid in $\langle N, 0, + \rangle$, and showed the completeness of the usual known axioms (commutativity, associativity of the sum and the zero axiom), and gives a very nice presentation of the topic in [Hen77].

Wilkie was the first to show with a proof theoretic approach in [Wil81] that even with 1 and the new axioms that come with 1, there are true identities that are not provable, like

$$(A^u + B^u)^v \cdot (C^v + D^v)^u = (A^v + B^v)^u \cdot (C^u + D^u)^v,$$

where $A = (x + 1)$, $B = (x^2 + x + 1)$, $C = (x^3 + 1)$, $D = (x^4 + x^2 + 1)$.

---

[25] Actually, he conjectured something stronger, namely that $\mathcal{E}$ is complete for $\langle N, Ack(n, \_, \_) \rangle$, the natural numbers equipped with a family of generalized binary operators $Ack(n, \_, \_)$ that extend the usual sum $+$, product $\cdot$ and exponentiation $\uparrow$ operators. In Tarski's definition, $Ack(0, \_, \_)$ is the sum, $Ack(1, \_, \_)$ is multiplication, $Ack(2, \_, \_)$ is exponentiation (for the other cases see for example [Rog88]).

[26] He also showed that there are no nontrivial equations for $\langle N, Ack(n, \_, \_) \rangle$ if $n > 2$.

Gurevič later gave a simpler model theoretic proof of this fact [Gur85] and showed finally that for the valid equalities of $\langle N, 1, +, \cdot, \uparrow \rangle$ there is no finite axiomatization, by means of an indexed family of equalities that generalize Wilkie's identity [Gur90]. Nonetheless, equality in all these structures, even if not finitely axiomatizable, is decidable [Mac81, Gur85].[27]

As often happens in number theory, these last results use far more complex tools than simple arithmetic reasoning, as in the case of [HR84], where Nevanlinna theory is used to identify a subclass of numerical expressions for which the usual axioms for $+$, $\cdot$, $\uparrow$ and 1 are complete.

This remarkable analogy between isomorphisms and equalities in number theory works quite well for the first-order case, even if it seems quite hard to find an extension to higher-order languages.

Other classes of categories

More recent results by Soloviev also provide a complete axiomatization of the isomorphisms of types in Symmetric Monoidal Closed Categories. It suffices to drop the nonlinear axioms (axioms 4 and 6 of Table 1.3) of the theory for ccc's [Sol93].

1.8.4 Isomorphisms in logic

Since the seminal work by Howard that introduced it, the *Curry-Howard isomorphism* has been one of the central themes in the theory of functional programming, as it is a bridge that allows to carry known results back and forth from one field to the other. It essentially says that typed λ-calculus can be seen as a system of notation for proofs in Intuitionistic Logic (obviously, different logical systems correspond to different calculi), in such a way that a given proposition A can be proved in the logical system $\mathcal{L}$ if and only if one can find a typed term $M : A$ in the correspondent calculus.

There is an exact correspondence, for example, between $\lambda^1 \beta \eta \pi *$ and the Intuitionistic Positive Calculus over the connectives **True**, $\wedge$, $\Rightarrow$, or $IPC(\textbf{True}, \wedge, \Rightarrow)$, when one reads **T**, $\times$, $\rightarrow$ respectively as **True**, $\wedge$, $\Rightarrow$.

Isomorphic types then come to correspond to propositions that are *strongly equivalent* in the terminology of [AB91, Mar92]:

Definition 1.8.5 (Strongly equivalent propositions) *Two propositions A and B are strongly equivalent if*

- $A \Longleftrightarrow B$ *(i.e., there exist derivations $f{:}A \Rightarrow B$ and $g{:}B \Rightarrow A$)*

---

[27]For the curious, here is how it works: from the size of the equation that has to be verified, it is possible to derive an upper bound; if the two sides coincide for all values of the variables up to this upper bound, then they coincide everywhere.

- *the composition $g \circ f$ of $f{:}A{\Rightarrow}B$ and $g{:}B{\Rightarrow}A$ by (CUT)*

$$\frac{\dfrac{f}{A{\Rightarrow}B} \qquad \dfrac{g}{B{\Rightarrow}A}}{A{\vdash}A} \quad (CUT)$$

reduces, after normalization[28], to the Axiom $\dfrac{\quad}{A{\vdash}A}$, and vice-versa.

Then the problem of finding all isomorphic types for a given typed λ-calculus corresponds to characterizing the *strongly equivalent* propositions in the associated logical system. This characterization can be fruitfully used to retrieve lemmas with equivalent proofs in the libraries of an automated theorem prover.

1.9   Isomorphisms and the lambda calculus

As we have seen in 1.8.3, the study of types that are isomorphic in all models of our typed λ-calculi can be seen as the study of those types for which there exist *definable* invertible conversion functions.

This suggests a different technique to study isomorphisms of types. Instead of looking at particular categories as done in Soloviev's work, we may focus on the invertible λ-terms that provide the conversion functions between isomorphic types. By a deep analysis of their structure, it is possible to gain enough information to clearly characterize all the isomorphisms that they can prove. In the following, we briefly review the basic notions and results that are fundamental to this approach to isomorphisms.

⚠ 1.9.1   Isomorphisms and invertibility

Long before the development of the recent interest on isomorphic types, there has been a series of very detailed studies of such conversion functions in the framework of the untyped λ-calculus, the most notable one being due to Dezani-Ciancaglini [Dez76], who proved that in the untyped case invertible terms have a very simple shape:

---

[28]Normalization of a proof can be something more complex of the usual cut-elimination [GLT90]: for example, we can perform on proofs the analogous of η-reductions in λ-calculus, which amounts to simplification to basic assumptions.

Theorem 1.9.1 (Invertible terms of λ-calculus) *Let M be an untyped term possessing normal form. Then M is invertible iff M is a finite hereditary permutation (f.h.p.).*

Finite hereditary permutations are defined inductively as follows.

Definition 1.9.2 (Finite hereditary permutations, f.h.p.)
An untyped λ term M is a finite hereditary permutation iff

- $M = \lambda x.x$, *or*

- $M = \lambda z.\lambda v_1 \cdots v_n.z P_1 \cdots P_n$ *and there exists a permutation $\sigma : n \to n$, such that $\lambda x_i.P_{\sigma(i)}$ is a finite hereditary permutation for all $1 \le i \le n$.*

Thus $\lambda z.\lambda x_1.\lambda x_2.z x_2 x_1$ and $\lambda z.\lambda x_1.\lambda x_2.z x_2 \lambda x_3.\lambda x_4.x_1 x_4 x_3$ are f.h.p.'s.

F.h.p.'s can also be tidily described in terms of Böhm-trees. Recall that a **Böhm-tree** of a term M is (informally) given by:

BT(M) $= \Omega$ if M has no head normal form

$$BT(M) = \quad \lambda x_1 \cdots x_n.y \qquad \text{if } M =_\beta \lambda x_1 \cdots x_n.y M_1 \cdots M_p$$
$$/\cdots\backslash$$
$$BT(M_1) \cdots BT(M_p)$$

Recall also that BT(M) is finite and Ω-free iff M has a normal form. Then one may look at f.h.p.'s as Böhm-trees, as follows:

$$\lambda z \overrightarrow{x}.z$$
$$/ \cdots \backslash$$
$$\lambda \overrightarrow{y_1}.x_{\sigma(1)} \qquad \cdots \qquad \lambda \overrightarrow{y_n}.x_{\sigma(1)}$$
$$\vdots \qquad\qquad\qquad \vdots$$

and so on, up to a finite depth (note that $\overrightarrow{y_i}$ may be an empty string of variables). Clearly f.h.p.'s are closed terms and they possess normal form. In particular, exactly the abstracted variables at level $n+1$ appear at level $n+2$, modulo some permutation of the order (note the special case of z at level 0).

These terms are all typable and hence are exactly the invertible terms of $\lambda^1 \beta \eta$.

Remark 1.9.3 *One may easily show that the f.h.p.'s are typable terms. (Hint: Just follow the inductive definition and give z, for instance, type $A_1 \to (A_2 \cdots \to B)$, where the A_i's are the types of the $N_{\sigma(i)}$.) By the usual abuse of language we may then speak of typed f.h.p.'s.*

Recall now that all typed terms possess a (unique) normal form (see [Bar84]). As we now need an interplay between typed and type-free terms, we are going to be more explicit about which sort of terms we are dealing with, when needed. Let M be a typed λ-term. We write e(M) for the **erasure** of M, i.e., for M with all type labels on variables erased.

Remark 1.9.4 *Observe that the erasures of all axioms and rules of the typed lambda calculus are themselves axioms and rules of the untyped lambda calculus. Then, in particular, if M and N are terms of $\lambda^1\beta\eta$ and $\lambda^1\beta\eta\vdash M = N$, one has $\lambda_{\beta\eta}\vdash e(M) = e(N)$.*

Theorem 1.9.5 *If $M : A \to B$ and $N : B \to A$ are invertible terms in $\lambda^1\beta\eta$, then e(M) and e(N) are f.h.p.'s.*

Proof. $e(N\ M) = e(N) \circ e(M)$, and hence, by the remark, $\lambda_{\beta\eta} \vdash e(M) \circ e(N) = e(I_\sigma) = I$ and $\lambda_{\beta\eta} \vdash e(N) \circ e(M) = e(I_\sigma) = I$. Thus by Theorem 1.9.1, e(M) and e(N) are f.h.p.'s. $\square$

Given this simple syntactic characterization, it is possible to proceed inductively on the structure of the terms to prove completeness of the axiom **Swap** for $\cong_d$ in the calculus $\lambda^1\beta\eta$(hence for $\cong$, due to Theorem 1.8.3), as is done in [BL85] (here Chapter 4).

This technique, unlike the semantic one, extends smoothly to the second-order case, the only real difficulty being the characterization of invertible terms. In the case of $\lambda^2\beta\eta$, this is done easily in [BL85], always using Dezani's result, and one gets that the invertible terms are the 2-f.h.p.'s defined as follows.

Definition 1.9.6 (Second-order f.h.p.) *A second-order term M of $\lambda^2\beta\eta$ is a second-order finite hereditary permutation (2-f.h.p.) iff*

- *$M = \lambda x.x$, or*

- *$M = \lambda z.\lambda v_1 \cdots v_n.z P_1 \cdots P_n$ and there exists a permutation $\sigma : n \to n$, such that for all $1 \leq i \leq n$*

    if $\lambda v_i = \lambda x_i : C$ then $\lambda x_i : C.P_{\sigma(i)}$ is a 2-f.h.p.
    if $\lambda v_i = \lambda X_i$ then $P_{\sigma(i)}$ is X_i.

Theorem 1.9.7 *2-f.h.p.'s are all and the only invertible terms of $\lambda^2\beta\eta$.*

Proof. By interpretation in the untyped calculus. See [BL85], Lemma 2.4 and Theorem 2.5. $\square$

This allows to show completeness of Th^2 for isomorphisms of $\lambda^2 \beta \eta$.

In the case of the calculi with constants, like $\lambda^1 \beta \eta \pi *$ or $\lambda^2 \beta \eta \pi *$, though, such a simple syntactic characterization is not already available. Actually, in the previous chapter only a subset of invertible terms is characterized, which is sufficient for the purposes of the completeness proof for $Th^1_{\times T}$. The guideline for the proof is to try to deal with the complexities risen by the different term constructors one at a time, in order to achieve a sort of factorization of the invertibility problem for the full calculus into the invertibility problem for a more manageable subclass of terms. The theory suggests that the type **T** is redundant and that the products in a type can be always pulled out of the other type-constructors, while still remaining in the class of isomorphic types. So the completeness proof is not as direct as it was in [BL85]; it needs some intermediate steps, very similar to those we will see here, but still stays rather simple, as Dezani's theorem can handle a relevant part of the complexity of the proof.

The second-order case with constants of $Th^2_{\times T}$ treated in this work is significantly more complex both than the pure second-order case and than the first-order case with constants. Here too we can give a characterization for a subclass of the invertible terms that is suitable for the purpose of proving the completeness of $Th^2_{\times T}$,[29] but the combination of the two extensions practically forces us to rebuild almost all the complex combinatorial proof of the original Theorem by Dezani, which can no longer be used simply as a tool out of the box.

This approach to isomorphisms via invertibility is particularly interesting because it does not depend on the existence of a special model like the category of natural numbers. Also, notice that given the types which are isomorphic in every models of the calculus, there is, in principle, no reason to believe that there is a uniform way to witness these isomorphisms. Nevertheless it turns out that our proofs of these results are based on a simple axiomatization of type equations and the notion of provable isomorphisms (those representable by closed terms of the lambda calculus). Moreover any proof of the equality of two types can be used to generate an uniform isomorphism between the types (which holds in every model).

1.9.2 The theories of isomorphisms for typed λ-calculi

As we have briefly seen here, the problems of characterizing isomorphic types, isomorphic objects and strongly equivalent propositions is really just one problem, as the solution to one of these is the solution to all the others. It is time for a summary of the known result: we present in Table 1.4 the

---

[29]For a generic invertible term we can give only a procedure to verify its invertibility and to build its inverse.

known theories of isomorphic types for typed λ-calculi, using a notation that is consistent with the names of the calculi.

So, for example, $Th^1_{\times \mathbf{T}}$ characterizes the isomorphic types for $\lambda^1 \beta \eta \pi *$: the number tells us that it is a first-order typed calculus, while the subscripts tell us that it has surjective pairing and unit type. Table 1.5 gives an overview of the connections between calculi, categories, proof systems and the associated theories, providing also bibliographical pointers to the published proofs of completeness of the theories.

The theory of isomorphic types for the second-order calculus, $Th^2_{\times \mathbf{T}}$ in Table 1.4, subsumes all the previously known ones for explicitly typed calculi (the proofs are rather more complex than the previous ones, too).

These theories are sound and complete, in the sense that

Definition 1.9.8 (Soundness, completeness) *We say that an equational theory Th is a* sound *theory of isomorphisms for a calculus (resp. class of categories, logical system) if*

$$\forall A, B \; Th \vdash A = B \Rightarrow A \cong B.$$

Respectively, an equational theory Th is a complete *theory of isomorphisms for a calculus (resp. class of categories, logical system) if*

$$A \cong B \Rightarrow \forall A, B \; Th \vdash A = B.$$

As can be expected, the soundness property is quite easy to show, while the completeness is a much harder property, and there are interestingly different techniques that can be used to establish it.

⚠ **1.9.3   Soundness**

For each of the different theories, it is possible to prove soundness either in a category-theoretic way (the axioms of $Th^1_{\times \mathbf{T}}$ are valid isomorphisms in every ccc: it is an easy exercise in elementary category theory) or by proof-theoretic techniques, but surely the easiest and more uniform way to soundness is by providing invertible terms of the corresponding typed calculi. The following proof actually provides us with soundness not only for $Th^2_{\times \mathbf{T}}$, but also for all the other theories. It suffices to see the corresponding invertible terms as embedded in the appropriate calculus. For example, $\lambda x : A \to (B \to C).\lambda y : B.\lambda z : A.xzy$ proves $A \to (B \to C) = B \to (A \to C)$ in $\lambda^1 \beta \eta$ if seen as a term of $\lambda^1 \beta \eta$, or in $\lambda^1 \beta \eta \pi *$ if seen as a term of $\lambda^1 \beta \eta \pi *$ and so on.

---

[30]Provided that X is free for Y in A, and that $Y \notin FTV(A)$

[31]Provided that $X \notin FTV(A)$

(**Swap**) $A \to (B \to C) = B \to (A \to C)$ $\bigr\} Th^1$

1. $A \times B = B \times A$
2. $A \times (B \times C) = (A \times B) \times C$
3. $(A \times B) \to C = A \to (B \to C)$
4. $A \to (B \times C) = (A \to B) \times (A \to C)$ $\bigr\} Th^1_{\times \mathbf{T}}$
5. $A \times \mathbf{T} = A$
6. $A \to \mathbf{T} = \mathbf{T}$
7. $\mathbf{T} \to A = A$

$\bigr\} Th^2_{\times \mathbf{T}}$

$\bigr\} - 10, 11 = Th^{ML}$

8. $\forall X.\forall Y.A = \forall Y.\forall X.A$
9. $\forall X.A = \forall Y.A[Y/X]$ ^{30}
10. $\forall X.(A \to B) = A \to \forall X.B$ ^{31}

$\bigr\} + \mathbf{Swap} = Th^2$

11. $\forall X.A \times B = \forall X.A \times \forall X.B$
12. $\forall X.\mathbf{T} = \mathbf{T}$

(**Split**) $\forall X.A \times B = \forall X.\forall Y.A \times (B[Y/X])$

A, B, C can be arbitrary types and $\mathbf{T}$ is a constant for the unit type. Notice that axiom **Swap** of Th^1 is provable in $Th^1_{\times \mathbf{T}}$ by axioms 1 and 3.

Table 1.4: The theories of valid isomorphisms for explicitly typed languages.

λ-Calculus	Category	Logical System	Theory	Authors
$\lambda^1\beta\eta$		$IPC(\Rightarrow)$	Th^1	([Mar72]), [BL85]
$\lambda^1\beta\eta*$		$IPC(\mathbf{True},\Rightarrow)$	$Th^1_{\mathbf{T}}$	Chapter 4
$\lambda^1\beta\eta\pi$	CC	$IPC(\wedge,\Rightarrow)$	$Th^1_{\times}$	Chapter 4
$\lambda^1\beta\eta\pi*$	CCC	$IPC(\mathbf{True},\wedge,\Rightarrow)$	$Th^1_{\times\mathbf{T}}$	[Sol83], [BDCL92]
				Chapter 4
$\lambda^2\beta\eta$		$IPC(\forall,\Rightarrow)$	Th^2	[BL85]
$\lambda^2\beta\eta*$		$IPC(\forall,\mathbf{True},\Rightarrow)$	$Th^2_{\mathbf{T}}$	Chapter 5
$\lambda^2\beta\eta\pi$		$IPC(\forall,\wedge,\Rightarrow)$	$Th^2_{\times}$	Chapter 5
$\lambda^2\beta\eta\pi*$		$IPC(\forall,\mathbf{True},\wedge,\Rightarrow)$	$Th^2_{\times\mathbf{T}}$	[DC94] , Chapter 5
Core-ML			Th^{ML}	Chapter 6

Table 1.5: Isomorphisms of types, objects and formulae.

Theorem 1.9.9 (Soundness) $Th^2_{\times T} \vdash A = B \Rightarrow A \cong B.$

Proof. By Theorem 1.8.3, it is enough to show that $Th^2_{\times T} \vdash A = B \Rightarrow A \cong_d B.$

For this purpose, we give the terms associated to each axiom and rule. As $Th^2_{\times T}$ is a theory of equality, one has first to observe that the usual axioms and inference rules yield and preserve provable isomorphisms:

- $\lambda x : A.x$ proves $A = A$;

- if M, with inverse N, proves $A = B$, then N proves $B = A$;

- if an invertible M proves $A = B$ and an invertible N proves $B = C$, then the term $N \circ M = \lambda x : A.N(Mx)$, which is clearly invertible, proves $A = C$;

- if an invertible term M proves $A = B$ and an invertible term N proves $C = D$, then the invertible term $\lambda x : A \times C.\langle M(p_1 x), N(p_2 x)\rangle$ proves $A \times C = B \times D$;

- if an invertible M proves $A = B$ and an invertible N proves $C = D$, then $\lambda y : A \to C.\lambda x : B.N(y(M^{-1}x))$, where M^{-1} is the inverse of M, proves $A \to C = B \to D$ and it is invertible (take $\lambda y : B \to D.\lambda x : A.N^{-1}(y(Mx))$);

- if an invertible M proves $A = B$, then $\lambda x : \forall X.A.\lambda X.M(x[X])$ proves $\forall X.A = \forall X.B$ and it is invertible (take $\lambda y : \forall X.B.\lambda X..M^{-1}(y[X]))$.

We next check the proper axioms:

1. $A \to (B \to C) = B \to (A \to C)$ is proved by

   $\lambda x : A \to (B \to C).\lambda y : B.\lambda z : A.xzy$, which is invertible;

2. $A \times B = B \times A$ is proved by

   $\lambda x : A \times B.\langle p_2 x, p_1 x\rangle$, which is invertible;

3. $A \times (B \times C) = (A \times B) \times C$ is proved by

   $\lambda x : A \times (B \times C).\langle\langle p_1 x, p_1(p_2 x)\rangle, p_2(p_2 x)\rangle$, which is invertible;

4. $(A \times B) \to C = A \to (B \to C)$

   is proved by $\lambda z : (A \times B) \to C.\lambda x : A.\lambda y : B.z\langle x, y\rangle$

   with inverse $\lambda z : A \to (B \to C).\lambda x : A \times B.z(p_1 x)(p_2 x)$;

5. $A \to (B \times C) = (A \to B) \times (A \to C)$

   is proved by $\lambda z : A \to (B \times C).\langle \lambda x : A.(\mathrm{p}_1(zx)), \lambda x : A.(\mathrm{p}_2(zx))\rangle$

   with inverse $\lambda z : (A \to B) \times (A \to C).\lambda x : A.\langle (\mathrm{p}_1 z)x, (\mathrm{p}_2 z)x\rangle$;

6. $\forall X.\forall Y.A = \forall Y.\forall X.A$

   is proved by $\lambda x : (\forall X.\forall Y.A).\lambda Y.\lambda X.((x[X])[Y])$

   with inverse $\lambda y : (\forall Y.\forall X.A).\lambda X.\lambda Y.((y[Y])[X])$;

7. $\forall X.A = \forall Y.A[Y/X]$

   is proved by $\lambda x : (\forall X.A).\lambda Y.(x[Y])$

   with inverse $\lambda y : (\forall Y.A[Y/X]).\lambda X.(y[X])$, provided that X is free for Y in A and Y is not free in A;

8. $\forall X.(A \to B) = A \to \forall X.B$

   is proved by $\lambda x : (\forall X.(A \to B)).\lambda y : A.\lambda X.(x[X])y$

   with inverse $\lambda z : (A \to \forall X.B).\lambda X.\lambda w : A.(zw)[X]$, provided that X is not free in A;

9. $\forall X.A \times B = \forall X.A \times \forall X.B$

   is proved by $\lambda x : (\forall X.A \times B).\langle \lambda X.(\mathrm{p}_1(x[X])), \lambda X.(\mathrm{p}_2(x[X]))\rangle$

   with inverse $\lambda y : (\forall X.A \times \forall X.B).\lambda X.\langle (\mathrm{p}_1 y)[X], (\mathrm{p}_2 y)[X]\rangle$;

10. $A \times \mathbf{T} = A$ is proved by $\lambda x.\mathrm{p}_1 x$ with inverse $\lambda x : A.\langle x, * \rangle$ (to check invertibility, notice that in $\lambda w : A \times \mathbf{T}.\langle \mathrm{p}_1 w, * \rangle$ we have $* = \mathrm{p}_2 w$ by equality *(top)*);

11. $A \to \mathbf{T} = \mathbf{T}$

    is proved by $rep((A \to \mathbf{T}) \to \mathbf{T}) = \lambda x : (A \to \mathbf{T}).*$

    with inverse $rep(\mathbf{T} \to (A \to \mathbf{T})) = \lambda y : \mathbf{T}.\lambda x : A.*$;

12. $\mathbf{T} \to A = A$ is proved by $\lambda x : (\mathbf{T} \to A).x*$ with inverse $\lambda y : A.\lambda w : \mathbf{T}.y$;

13. $\forall X.\mathbf{T} = \mathbf{T}$ is proved by $\lambda x : \forall X.\mathbf{T}.*$ with inverse $\lambda y : \mathbf{T}.\lambda X.*$.

    Here for the difficult side of the equality, notice that if $M : \forall X.\mathbf{T}$ then $M[X] = *$, by equality *(top)*, so that, for X a fresh type variable, $\lambda X.M[X] = \lambda X.*$ by rule ξ and finally by η we get $M = \lambda X.M[X] = \lambda X.*$. Hence

$$(\lambda y : \mathbf{T}.\lambda X.*) \circ (\lambda x : \forall X.\mathbf{T}.*) =$$
$$= \lambda w : \forall X.\mathbf{T}.(\lambda y : \mathbf{T}.\lambda X.*)((\lambda x : \forall X.\mathbf{T}.*)w)$$
$$= \lambda w : \forall X.\mathbf{T}.(\lambda y : \mathbf{T}.\lambda X.*)*$$
$$= \lambda w : \forall X.\mathbf{T}.\lambda X.*$$
$$= \lambda w : \forall X.\mathbf{T}.w$$
$$= \mathrm{I}_{\forall X.\mathbf{T}}$$

□

Chapter 2

Confluence Results

2.1   Introduction

In the λ-calculus there are plenty of programs (or λ-terms) where it is possible to apply the reduction rule β, introduced in 1.3.4, in more than one position. The λ-calculus does not specify a unique evaluation order, or *reduction strategy*, that associates to each program or term a unique position where the reduction will proceed.

This is usually one of the distinguishing features of a *calculus* with respect to a programming language: a programming language can often be seen as a particular instance of a calculus where we specify an evaluation order, hence the slogan

"programming language = calculus + reduction strategy".

In a calculus we are interested in the theoretical study of those properties that are independent of the particular evaluation order chosen. In these chapters, we focus on two of these general properties: *confluence* and *normalization*. To do this properly, let us recall some notions from rewriting theory.

Notation 2.1.1 (reduction and normal forms) *As usual, $\rightarrow$ will denote one-step reduction, while $\rightarrow_=$ is the reflexive closure of $\rightarrow$, and $\overset{*}{\rightarrow}$ is the reflexive transitive closure of $\rightarrow$. A normal form (n.f.) is a term that cannot be reduced any further, i.e., from which no $\rightarrow$ step can leave.*

A reduction system is often given by means of a set R of *reduction rules*, that may be made explicit writing $\overset{R}{\longrightarrow}$, $\overset{R}{\longrightarrow}_=$ and $\overset{R}{\rightarrow}^*$. When we talk about *reduction*, we emphasize the *operational* view of a calculus, where we are interested in how to *reduce* terms in order to get their *values*, whatever they are defined to be. On the other hand, when we want to emphasize a more logical view of a calculus, we focus on its associated *equality*: we want to be able to describe precisely when two terms are equal, i.e., they have for us the same *meaning*, whatever that meaning is defined to be.

These two views of a calculus are usually interchangeable, in the following sense. If we are given a calculus with a notion of reduction $\rightarrow$, then the associated equality is defined as the smallest equivalence relation that contains $\rightarrow$, and can be usually[1] built as the *reflexive transitive and symmetric closure of $\rightarrow$*: two terms are equal if they are connected by a sequence of reduction steps, not necessarily all oriented in the same direction. We say

---

[1] Usually, but not always. For conditional notions of reduction like the expansion rules we will meet later on, this construction is not correct.

that this is the equality *generated by* the reduction. If we take as meaning of a term its normal form w.r.t. →, and if the reduction system has some good properties (confluence, in particular), this precisely captures the intuition that two terms are equal if and only if they have the same meaning. Conversely, if we are given a notion of equality $\mathcal{E}$ on terms, there are many ways to define an associated notion of reduction. The only requirement is that the notion of reduction → that we choose generates $\mathcal{E}$. It is here that it is quite important to state some good properties of a notion of reduction, like confluence and normalization: when we are given an equality on terms and asked to choose an associated reduction, these properties will serve as necessary guidelines.

The confluence property says that any two reduction sequences starting from the same term have a common reduct (i.e., can be prolonged in a way that allows to reach a same term, Fig. 2.1 b). It is easy to see that confluence is equivalent to the Church-Rosser property, that says that whenever two terms are equal, they have a common reduct, Fig. 2.1 a; for this reason it is customary to refer indifferently to one or the other as the CR property. Of technical interest is also the weak confluence property, WCR, where the existence of a common reduct is assured only when the two reduction sequence starting form the same term have length one, Fig. 2.1 c.

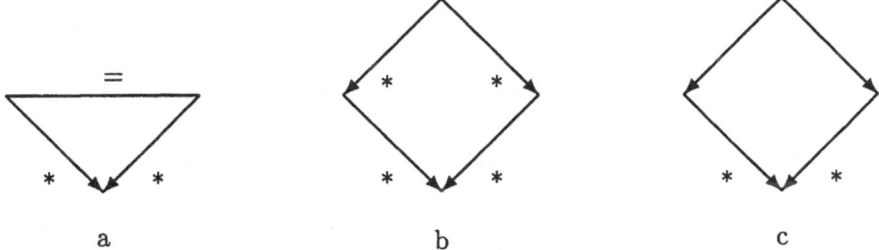

Figure 2.1: Church-Rosser, Confluence and Weak Confluence properties.

On the other hand, normalization is concerned with the possibility of reaching in a finite number of steps a normal form. If for any term there exists a sequence of reduction steps that leads to a normal form one says that the system is weakly normalizing, or WN. If a normal form is reached no matter what sequence of reduction is used, then the system is strongly normalizing, or SN.

Of course, these properties are usually more general and more difficult to prove for a calculus than in the case of a specific programming language

based on such a calculus. For example, if we can prove that any possible reduction in a calculus terminates after a finite number of steps (i.e., that the
calculus is strongly normalizing), then we have proven that all languages
based on this calculus are strongly normalizing. In the case of a programming language, confluence is in general a trivial property, since there are
never two different evaluation sequences starting from the same program.[2]
Nevertheless, suppose that two different programming languages have the
same underlying calculus and share with the calculus the notion of normal
form. Then, if the calculus is confluent, we know that the normal forms of
a same program computed in these two different programming languages
are the same.[3] This fact helps the language designer to choose more easily between different programming languages, or even allows him to leave
unspecified part of the reduction strategy, by telling him that whatever
strategy the implementor will choose, since it will not affect the result of
any program. This will in turn leave the implementor more space for de-
signing efficient execution mechanisms for the programming language.

In this and the following chapter we will develop a basic theory of reduction for the typed calculi that form the basis of our study of isomorphic
types in λ-calculus. We provide a confluent notion of reduction that is also
weakly normalizing. This will allow us in the rest of the book to work
on terms in normal form, knowing that such normal form is unique. As
will become evident from the reading of the survey, confluence and normalization for the typed λ-calculi we study here are results that are far from
being simple variations of standard known techniques: new tools need to
be developed. Most of the material presented in this chapter and in the
following one comes from joint work with Pierre-Louis Curien, summarized
in [CDC91], but we present also a general overview of the different proof
techniques and results developed before and after [CDC91] appeared. For a
more detailed presentation of term-rewriting systems and their properties,
we refer the interested reader to [GD80, DJ90, Hue80].

---

[2]This is the case in deterministic sequential languages that have a well-defined semantics: C is a notable exception, as its lack of a formal semantics opens the door to
syntactically valid expressions, like the function call f(i++,i--), whose behavior, as the
cognoscenti will know, is not specified.

[3]These conditions are not so far-fetched as they might seem: in particular, on base
types, like integers, reals, strings and the like, almost all programming languages share
the same notion of normal form.

2.2 Extensionality

Over the past years there has been a growing interest in the properties of
λ-calculus extended with various different type-constructors, in particular
pairs and sums, used to represent common data types. For these type-
constructors it is customary to provide a set of equalities that are then
turned into computation rules. This is the case, for example, of the elimination rules for pairs:

$$(\pi_1) \quad \mathrm{p}_1(\langle M, N\rangle) \quad = \quad M \qquad (\pi_2) \quad \mathrm{p}_2(\langle M, N\rangle) \quad = \quad N$$

They tell us how to operationally *compute* with objects of these types. If we
have a pair $\langle M, N\rangle$, then we can decompose it to access its first or second
component.

There is anyway something else that one likes to do with λ-calculus,
besides using λ-terms as programs to be computed. One would like to
reason about programs, to prove that they enjoy certain properties. Here
is where extensional equalities come into play. In the case of functions, for
example, since the only operational way to *use* a function is to apply it
to an argument, we do not really want to consider a term M of function
type different from the term $\lambda x.Mx$ where x does not occur free in M:
both terms, when applied to an argument N, give the same result MN.
Similarly for pairs, the only operational way to *use* a pair is by projecting
out the first or the second component, so we do not want to consider a term
M of product type different from the term $\langle \mathrm{p}_1(M), \mathrm{p}_2(M)\rangle$: the result of
accessing any of these two terms via a first or second projection is the same
term $\mathrm{p}_1(M)$ or $\mathrm{p}_2(M)$.

These facts can be incorporated in the calculus in the form of equalities,
which one can read in at least two different ways:

- *an operational way*: these equalities just state possible *optimizations*
  of a program. Since a term $\langle \mathrm{p}_1(M), \mathrm{p}_2(M)\rangle$ is more complex then M
  but behaves the same way, it is convenient to replace all its occur-
  rences by M, as this transformation will yield a more efficient and
  smaller, but equivalent, program. Similarly, we will replace every
  occurrence of $\lambda x.Mx$ by M.

- *a theoretical way*: these equalities state a relation between a program
  and its type. They just tell us that whenever a term M has a functional type, then it must really be a function, built by λ-abstraction,
  so we ought to replace it by $\lambda x.Mx$ if it is not already a function. Similarly, a term M of product type has to be really a pair, built via the
  pair constructor, or otherwise it must be replaced by $\langle \mathrm{p}_1(M), \mathrm{p}_2(M)\rangle$.

As we will briefly see in the survey, over the past twenty years a lot of research activity has focused both on the operational and the theoretical readings of these equalities.

2.2.1   Survey

Due to the deep connections between λ-calculus, proof theory and category theory, works on extensional equalities have appeared with different motivations in all these fields.

By far, the best-known extensional equality is the η axiom that we informally introduced above, written in the λ-calculus formalism as

$$(\eta) \qquad \lambda x.Mx = M \qquad \text{provided } x \text{ is not free in } M$$

This axiom, also known as *extensionality*, has traditionally been turned into a reduction, carrying the same name, by orienting the equality from left to right, interpreting operational equality as a *contraction*. Such an interpretation is well behaved as it preserves confluence [CF58].

In the early 70s, the attention was focusing on products and the extensional rule for pairs, called *surjective pairing*, which is the analog for product types of the usual η extensional rule.

$$(SP) \qquad \langle p_1(M), p_2(M) \rangle = M$$

Contractions

With the previous experience of the η rule, it is easy to understand how, at that time, most of the people thought that the right way to turn such an equality into a rewrite rule was also from left to right, as a contraction. But in 1980, J.W. Klop discovered [Klo80] that, if added to the usual confluent rewrite rules for pure λ-calculus, this interpretation of SP breaks confluence.[4]

Anyway, this first negative result was mitigated shortly after in [Pot81] for the simply *typed* λ-calculus with η and SP contractions, by providing a first proof of confluence and strong normalization, later simplified in different ways (see [Tro86] or [GLT90], for example). From then on, the contraction rule for SP was not considered harmful in a typed framework until the seminal work by Lambek and Scott [LS86]. There, the decision problem of the equational theory of Cartesian Closed Categories (ccc's) is solved using a particular typed λ-calculus equipped with not only η and SP equalities, but also with a special type **T** representing the *terminal object* of

---

[4]See [Bar84], p. 403-409 for a short history and references.

the ccc's.[5] This distinguished atomic type comes with a further extensional axiom asserting that there is exactly one term $*$ of type $\mathbf{T}$:

$$(Top) \qquad M = * \qquad \text{if } M : \mathbf{T}$$

Now, the type $\mathbf{T}$ has the bad property of destroying confluence, if the extensional equalities η and SP are turned into contraction rules. The following are the critical pairs that arise immediately, as first pointed out by Obtulowicz (see [LS86]):

$$\langle p_1(x), p_2(x) \rangle \to_{SP} x \qquad\qquad \langle p_1(x), p_2(x) \rangle \to_{SP} x$$
$$\downarrow \mathbf{T} \qquad\qquad\qquad\qquad\qquad \downarrow \mathbf{T}$$
$$\langle *, p_2(x) \rangle \qquad\qquad\qquad\qquad \langle p_1(x), * \rangle$$

$$(\lambda x : \mathbf{T}.Mx) : (\mathbf{T} \to A) \to_\eta M \qquad (\lambda x : A.Mx) : (A \to \mathbf{T}) \to_\eta M$$
$$\downarrow \mathbf{T} \qquad\qquad\qquad\qquad\qquad\qquad \downarrow \mathbf{T}$$
$$(\lambda x : \mathbf{T}.M*) : (\mathbf{T} \to A) \qquad\qquad (\lambda x : A.*) : (A \to \mathbf{T})$$

It is indeed possible, but not easy, to extend the contractive reduction system in order to recover confluence. A first step towards such a confluent system was taken by Poigné and Voss, who were not inspired by category theory but by the implementation of algebraic data types [PV87]. In their paper, they study a calculus that includes $\lambda^1\beta\eta\pi*$ and notice that to solve the previous critical pairs one needs to add an infinite number of reduction rules (that can be anyway finitely described). Then confluence of such an extended system can be proved by showing weak confluence and strong normalization. Unfortunately, the critical pair for $(\lambda x : A.Mx) : \mathbf{T} \to A$ is missing there, and the strong normalization proof is incomplete.

More recently, Curien and this author got interested in a polymorphic extension of $\lambda^1\beta\eta\pi*$ that arose in the study of the theory of object-oriented programming and of isomorphisms of types [CDC91]. In the framework of inheritance, the terminal type $\mathbf{T}$ has an additional flavor: it is a maximum type. Type inclusion is *not* invariant under isomorphisms, so that, say $A \times \mathbf{T}$ is a type greater than $A \times A'$ for any A', while the same is not true of A.[6] We give a complete (infinite) set of reduction rules for the calculus, which is

---

[5]This is the *Unit* type in languages like ML.

[6]Recently, L. Cardelli has proposed the following nice and simple exploitation of $\mathbf{T}$ as a maximum type. Consider the well-known inheritance [age;sex] less than [age]; encode [age] as $age \times \mathbf{T}$ and [age;sex] as $age \times (sex \times \mathbf{T})$. Then the desired subtyping obviously holds componentwise, by reflexivity and maximality, respectively.

proved confluent using just weak confluence, weak normalization and some
additional properties. Moreover, we can take profit of confluence to get
conservativity results in addition to decision results. Such conservativity
results are needed in the study of provable type isomorphisms. The results
are applied there in two related areas:

- Foundations of object-oriented programming. [CG90] focuses on the
  paradigmatic language $F_\leq$, a variant of second-order λ-calculus with
  a maximum type and bounded quantification: the equational theory
  considered consists of β, η (first and second-order) and the terminal
  type rule. Confluence of that system is shown via a translation to
  the polymorphic λ-calculus with a terminal type (what is called here-
  after $\lambda^2\beta\eta*$), and by using a general criterion allowing to transfer
  confluence in $\lambda^2\beta\eta*$ back to the source system.

- Study of isomorphisms of types. We will give in Chapters 4 and 5 an
  equational characterization of all type isomorphisms which are provable in the typed λ-calculus (respectively second-order λ-calculus)
  with pairs and terminal object ($\lambda^1\beta\eta\pi*$, respectively $\lambda^2\beta\eta\pi*$). It
  turns out that this characterization can be given quite easily if we
  are able to determine the structure of *invertible* terms, i.e. terms that
  possess an inverse w.r.t. the usual operation $\lambda x.\lambda y.\lambda z.(x(yz))$ of composition. The conservativity of equality in the extended calculus over
  the calculus without products and terminal objects allows us to reduce the problem to the invertibility in the simply typed (respectively
  second-order) λ-calculus.[7]

Expansions

Meanwhile, in the field of proof theory, Prawitz was suggesting [Pra71] to
turn these extensional equalities into *expansion* rules, rather than contractions. Building on such ideas, but motivated by the study of coherence
problems in category theory, Mints gives a first faulty proof that in the
typed framework *expansion rules*, if handled with care, are weakly normalizing and preserve confluence of the typed calculus [Min79].[8]

This idea of using expansion rules seems to have passed unnoticed for
a long time, even if the so called η-long normal forms were well known
and used in the study of higher-order unification problems [Hue76]. Only
in these last years there has been a renewed interest in expansion rules.

---

[7]Ultimately the problem is reduced to the invertibility in the untyped λ-calculus (see
[Bar84], Section 21.2), where invertible terms have the simple syntactic characterization
described in 1.9.2.

[8]The same idea is present in [Min77].

In recent work [Jay92], still motivated by category theoretic investigation, Jay explores a simply typed λ-calculus with just **T** and a natural number type **N** as base types, equipped with an induction combinator for terms of type **N**. He introduced expansion rules for η and SP that are exactly the same as the ones originally used by Mints, and in [JG92] this calculus is proved confluent and strongly normalizing. Category theory is also the motivation of Cubric [Cub92], who repaired the error in the original proof by Mints, showing confluence (by means of a careful study of residuals) and also weak normalization (but not strong normalization). An interesting divide-and-conquer approach is proposed in [Aka93], where confluence and strong normalization are shown in a modular way. In [Dou93], confluence is shown via the usual Newman's Lemma, and strong normalization by means of a variation of the reducibility proof based on introduction rather than elimination terms.

In [DCK93, DCK94b], Delia Kesner and the author use expansion rules to provide a confluent rewriting system for the typed λ-calculus with not only products and terminal object, but also sums and recursion. This result is derived from the confluence of a restricted system where recursion is bounded (recursive calls of infinite length are not allowed). This system is proved to be weakly confluent by an analysis of the critical pairs, and strongly normalizing by means of an appropriate *simulation* of any one-step reduction in the calculus with expansions by a non-empty reduction sequence in the calculus without expansions. It turns out that this result is powerful enough to prove directly also the confluence property.

This last very detailed work was the necessary basis of [DCK94a], where it is shown that expansion rules can be used to preserve modularity of confluence in combination of typed λ-calculi with algebraic rewriting systems and fixpoint combinators, that are results of much more general interest.

The interested reader is suggested to browse through the works listed in Figure 2.2, which provides a hopefully complete overview of the research done in this area.

The expansive approach is quite promising today, but in this chapter we choose to present the result obtained in [CDC91], as it provides a confluent and effectively weakly normalizing (thus decidable) rewriting systems for the full equational theory underlying cartesian closed categories *and* for polymorphic extensions of it. This work still provides the first solution to the decidability problem of the full equational theory of cartesian closed categories extended with polymorphic types, and it is the one that was

1970s: the first expansion

1971 Prawitz suggests to reverse η [Pra71]

1976 Huet uses $\beta\eta$-long normal forms for higher-order unification [Hue76]

1979 Mints reverses η and SP [Min79]

197- Many people suggest expansions: Martin-Löf, Meyer, Statman, etc.

1980s: the contraction

1980 Klop's counterexample to CR for λ+SP [Klo80]

1981 Pottinger shows CR for *typed* $\lambda\beta\eta$+SP [Pot81]

1986 Lambek and Scott, Obtulowicz: typed $\lambda\beta\eta$+SP+**T** is not CR [LS86]

1987 Poigné and Voss try completion for $\lambda\beta\eta$+SP+**T**+sums and recursion [PV87]

1991 Curien   and   Di   Cosmo: completion   for   *polymorphic* $\lambda\beta\eta$+SP+**T** [CDC91]

1990s: the second expansion

1991 Jay: SN for expansions+**T**+**N** [Jay92]

1992 Di Cosmo and Kesner: CR+SN for expansions+**T**+sums+weak extensional sums, CR with recursion [DCK93, DCK94b]

1992 Cubric: CR for expansions+**T** [Cub92]

1992 Ghani and Jay: CR+SN for expansions+**T**+**N** [JG92]

1992 Akama: SN+CR for expansions+**T** [Aka93]

1992 Dougherty: CR+SN for expansions+**T**+sums, CR with recursion [Dou93]

1993 Di Cosmo and Kesner: modularity of CR and SN for expansions + algebraic systems, of CR for recursion [DCK94a]

1993 Piperno and Ronchi Della Rocca use expansions to study polymorphic type inference [PRDR94]

Figure 2.2: A brief history of extensionality.

historically used for the study of isomorphisms of types.

2.3 Overview

Technically, we had to navigate between several pitfalls before we arrived to our solution for contractive extensional rules in [CDC91], which is presented here. We survey the main steps of this eventually safe trip in this section. Sections 2.4 and 2.5 are devoted to confluence and weak normalization, respectively. In Section 2.6 we state the decidability and conservativity results that follow quite obviously from confluence and weak normalization, and we put our work in perspective with the other approaches to decidability of the same theories that we are aware of. Section 2.7 is a brief conclusion, which applies as well to Chapter 3. We use the Knuth-Bendix procedure by hand to obtain locally confluent rewriting systems. We then shortly hint at a severe technical difficulty in adapting the standard strong normalization proofs which use the so-called reducibility method. They can be adapted to a subsystem only. From the confluence of this subsystem we get confluence of almost the whole system by a general criterion presenting an interest of its own. At this stage, only the second-order β-rule is left out, and it can be finally added with the help of Hindley-Rosen's Lemma. As for weak normalization, the ingredients developed for confluence give it for free for first-order systems, while for the second-order systems another splitting in subsystems and another adaptation of the standard strong normalization proofs are needed.

2.3.1 Weakly confluent reduction

The systems obtained by orienting the equalities of $=_{\beta^2 \eta^2 \pi *}$ and its restrictions are far from being even weakly confluent, due to a bad interaction between the rule *top* on one side and the rules η and SP on the other.[9] The point is that all terms of type $\mathbf{T}$ are identified (in particular, $x : \mathbf{T}$ and $*$ are identical), so that $\lambda x : \mathbf{T}.Mx$ and $\lambda x : \mathbf{T}.M*$ are "the same" term and must give rise to the same reductions. Since the first reduces to M, the second must reduce to M too. This fact actually shows up during the completion procedure. Let us consider the typical critical pairs that arise, say, for $\lambda^2 \beta \eta \pi *$: after the first "stage" we find the situation described in Figure 2.3.

---

[9]This observation seems to have been first made by A. Obtulowicz, cf. [LS86], exercise at page 88.

	M	M'	M''	New reduction from completion
$\eta - like$	$\lambda x : \mathbf{T}.Mx$	M	$\lambda x : \mathbf{T}.M*$	$\lambda x : \mathbf{T}.M* \longrightarrow M$ if $x \notin FV(M)$
	$\langle \mathrm{p}_1 M, \mathrm{p}_2 M \rangle$	M	$\langle \mathrm{p}_1 M, * \rangle$	$\langle \mathrm{p}_1 M, * \rangle \longrightarrow M$ if $M : A \times \mathbf{T}$
	$\langle \mathrm{p}_1 M, \mathrm{p}_2 M \rangle$	M	$\langle *, \mathrm{p}_2 M \rangle$	$\langle *, \mathrm{p}_2 M \rangle \longrightarrow M$ if $M : \mathbf{T} \times B$
$top - like$	$\lambda x : A.Mx$	M	$\lambda \mathrm{x}{:}A.*$	$M \longrightarrow \lambda x : A.*$ if $M : A \to \mathbf{T}$
	$\lambda X.M[X]$	M	$\lambda X.*$	$M \longrightarrow \lambda X.*$ if $M : \forall X.\mathbf{T}$

Figure 2.3: The critical pairs at the first stage of Knuth-Bendix completion. (M' is reached via η or SP; M'' via top)

The additional rules generated by completion can be divided in two groups: rules that behave like η (*eta-like*) and rules that behave like *top* (*top-like*). The former mimics the behavior of η and SP rules on terms that are known to be "the same terms as" η and SP redexes, as in the example we just considered above. The latter force to identify all the terms of type $A \to \mathbf{T}$ and $\forall A.\mathbf{T}$, and do pick up a canonical representative in the respective types. It turns out that a set of *eta-like* rules must be generated for each of all types isomorphic (in the categorical sense, see [BDCL92] and [DC94]) to $\mathbf{T}$. At stage n, the completion procedure on one side creates new rules to mimic η and SP on terms that are known to be "the same" as *eta-like* stage $n - 1$ redexes, and on the other side it discovers new "same" terms, following the pattern:

- if A is known to be isomorphic to $\mathbf{T}$ at stage $n - 1$, then $B \to A$ and $\forall X.A$ are isomorphic to $\mathbf{T}$ at stage n

- if A and B are known to be isomorphic to $\mathbf{T}$ at stage $n - 2$, then $A \times B$ is isomorphic to $\mathbf{T}$ at stage n.

These correspond to the well-known isomorphisms $\mathbf{T} \times \mathbf{T} \cong \mathbf{T}$, $A \to \mathbf{T} \cong \mathbf{T}$ and $\forall X.\mathbf{T} \cong \mathbf{T}$. (The isomorphism $\mathbf{T} \times \mathbf{T} \cong \mathbf{T}$ shows up only from the second stage on: consider the stage 1 *eta-like* redex $\langle *, \mathrm{p}_2 M \rangle$, and suppose $M : \mathbf{T} \times \mathbf{T}$. Then we reach M by the *eta-like* reduction, and $\langle *, * \rangle$ by *top*.)

The following notation will allow us to present in a compact formalism the resulting weakly confluent reduction system.

Definition 2.3.1 *Terminal types and canonical terms.*

1. $Iso(\mathbf{T})$ (the collection of types isomorphic to $\mathbf{T}$) is the set defined as follows:

    (a) $\mathbf{T} \in Iso(\mathbf{T})$
    (b) if $B \in Iso(\mathbf{T})$, then $A \to B \in Iso(\mathbf{T})$ for every type A
    (c) if $A \in Iso(\mathbf{T})$ and $B \in Iso(\mathbf{T})$, then $A \times B \in Iso(\mathbf{T})$
    (d) if $A \in Iso(\mathbf{T})$ and X is a type variable, then $\forall X.A \in Iso(\mathbf{T})$.

2. For each type $A \in Iso(\mathbf{T})$, the associated *canonical* representative $rep(A)$ is defined inductively as follows:

    (a) $rep(A \times B)$ is $\langle rep(A), rep(B) \rangle$ (c) $rep(\mathbf{T})$ is $*$
    (b) $rep(A \to B)$ is $\lambda x : A.rep(B)$ (d) $rep(\forall X.A)$ is $\lambda X.rep(A)$.

Definition 2.3.2 $\overset{\beta^2\eta^2\pi*}{\longrightarrow}$ *is the notion of reduction for $\lambda^2\beta\eta\pi*$ generated by orienting to the right the equalities β, η, π, SP, β^2 and η^2 in Definition 1.5.1 and adding the following rewriting rules, coming from completion:*

(*gentop*) $M{:}A \overset{\beta^2\eta^2\pi*}{\longrightarrow} rep(A)$ if $M : A$ and $A \in Iso(\mathbf{T})$ and M is not already $rep(A)$

(SP_{top}) $\langle rep(A), \mathrm{p}_2 M \rangle \overset{\beta^2\eta^2\pi*}{\longrightarrow} M$ if $M{:}A \times B$

(SP_{top}) $\langle \mathrm{p}_1 M, rep(B) \rangle \overset{\beta^2\eta^2\pi*}{\longrightarrow} M$ if $M{:}A \times B$

(η_{top}) $\lambda x : A.M\, rep(A) \overset{\beta^2\eta^2\pi*}{\longrightarrow} M$ if $A \in Iso(\mathbf{T})$ and $x \notin FV(M)$.

The notions of reduction for the simpler calculi can be defined as restrictions of $\overset{\beta^2\eta^2\pi*}{\longrightarrow}$. The notion of reduction for $\lambda^2\beta\eta*$, which we will note $\overset{\beta^2\eta^2*}{\longrightarrow}$, is the reduction induced on $\lambda^2\beta\eta*$ by $\overset{\beta^2\eta^2\pi*}{\longrightarrow}$, that is $\overset{\beta^2\eta^2\pi*}{\longrightarrow}$ without π, SP, and SP_{top}, as these rules cannot apply to terms of $\lambda^2\beta\eta*$. For the same reason, the clauses for product types in Definition 2.3.1 will never be used, so that actually only a restricted version of *gentop* is used in $\overset{\beta^2\eta^2*}{\longrightarrow}$. We shall still use *gentop* to name this restricted reduction, as the intended meaning will always be clear from the context. Similarly, $\overset{\beta\eta\pi*}{\longrightarrow}$ and $\overset{\beta\eta*}{\longrightarrow}$ are the reductions induced by $\overset{\beta^2\eta^2\pi*}{\longrightarrow}$ on $\lambda^1\beta\eta\pi*$ and $\lambda^1\beta\eta*$, with the appropriate restrictions of *gentop*.

It is now just a matter of an easy structural induction on the reductions to see that

Proposition 2.3.3 $\overset{\beta^2\eta^2\pi*}{\longrightarrow}$ *is weakly confluent (WCR).*

Proof. For every $M' \leftarrow M \longrightarrow M''$, we need to find a term M''' s.t. $M' \longrightarrow M''' \leftarrow M''$.

We proceed by induction on the derivation of $M \longrightarrow M'$. In the case $M' \leftarrow M \longrightarrow M''$ involves a reduction at the root of M, then we are in the presence of a critical pair, and the term M''' is given by the Knuth-Bendix completion construction above.

If both the reductions are done in subterms of M, i.e., if $M \longrightarrow M'$ and $M \longrightarrow M''$ are derived by a context closure rule, then we distinguish two cases:

- The reduction takes place in subterms that are not overlapping, like in the case

$$\langle M_1', M_2 \rangle \leftarrow \langle M_1, M_2 \rangle \longrightarrow \langle M_1, M_2'' \rangle.$$

Then we can close the diagram using the original reductions in the reversed order, like in

$$\langle M_1', M_2 \rangle \longrightarrow \langle M_1', M_2'' \rangle \leftarrow \langle M_1, M_2'' \rangle.$$

- The reductions take place in the same subterm M_1, like in the case

$$\lambda x.M_1' \leftarrow \lambda x.M_1 \longrightarrow \lambda x.M_1'',$$

so the last step of the derivation was a context closure rule. Then we can find by induction hypothesis a term M_1''' that completes the diagram $M_1' \leftarrow M_1 \longrightarrow M_1''$ via $M_1' \longrightarrow M_1''' \leftarrow M_1''$, and then (applying the same context closure rule) complete the original diagram, like in

$$\lambda x.M_1' \longrightarrow \lambda x.M_1''' \leftarrow \lambda x.M_1''.$$

$\square$

What about confluence then? We cannot use the standard Tait-Martin Löf "parallel reduction" technique, as the nonlinear rule SP may require more than one adjustment step, which cannot be parallelized. Specifically, suppose that M one-step reduces to M': then $\langle p_1 M, p_2 M \rangle$ reduces both to M and to $\langle p_1 M', p_2 M' \rangle$. The local confluence diagram can be completed on one side in one step to M', but on the other side one must go sequentially to $\langle p_1 M', p_2 M' \rangle$, where the lost SP redex is recreated, and then to M'. This is hardly parallel.

2.3.2 Investigating strong normalization

Another "obvious" approach to prove confluence is to attempt to show that these notions of reduction are strongly normalizing, as then one could apply the well known fact that $SN + WCR \Rightarrow CR$.[10] But here we face a serious problem: some of the new reduction rules, namely η_{top} and SP_{top}, prevent us from applying the usual reducibility techniques (see [GLT90], [LS86], [Tai67]), as we briefly sketch now.

All variations of the reducibility method require to show a key statement like "if $v[u/x] \in RED_V$ for all $u \in RED_U$, then $\lambda x.v \in RED_{U \to V}$", where RED_T is the set of reducible terms of type T, and where $RED_{U \to V}$ is the set of s:$U \to V$ s.t. $(su) \in RED_V$ for all $u \in RED_U$.

An auxiliary property which is available is that, whenever $(st) : T$, one has $(st) \in RED_T$ as soon as $s' \in RED_T$ for all s' which are one step reducts of (st).

So the proof of the key statement reduces to the proof that all one step reducts of $(\lambda x.v) u$ are reducible. Now, if v is $(v'*)$, then $(\lambda x.v)u$ can reduce to $(v'u)$ that is *not* $v[u/x] = v$, and we do not know if $(v'u)$ is reducible: this does not follow from any of the hypotheses we have at hand. A similar situation arises for SP_{top} when considering the corresponding lemma for pairs. (See the Remark 3.1.14 in Section 3.1.)

But the difficulty suggests a solution. The above example is problematic only if u is different from $*$, and this cannot happen if we restrict our attention to terms in *gentop* normal form (*gentop* n.f.). For this to work out we have to check that *gentop* normal forms are stable under reduction. Otherwise the problem could dynamically show up later in the reduction. Unfortunately the β^2 rule does not preserve *gentop* normal forms.

Example 2.3.4 Let A be a type with no occurrence of **T**. Then the second-order term $(\lambda X.\lambda x : X.\lambda y : (X \to A).yx)[\textbf{T}]$ is in *gentop* normal form, but its contractum $\lambda x : \textbf{T}.\lambda y : \textbf{T} \to A.yx$ is not, and reduces to $\lambda x : \textbf{T}.\lambda y : \textbf{T} \to A.y*$. $\square$

So we are forced to drop β^2. Summarizing, so far we have hopes for confluence in the system which is restricted in two ways: we work only with *gentop* normal forms and we have abandoned β^2. Indeed, we show that this restricted system is strongly normalizing (Section 3.1) andthus confluent (the proof of local confluence is easily adapted to the subsystem). Then we lift the confluence result to the system $\xrightarrow{\beta\eta^2\pi*}$, as we will note the notion of reduction induced on $\lambda^2\beta\eta\pi*$ by $\xrightarrow{\beta^2\eta^2\pi*}$ less β^2 (see next subsection).

---

[10]Known as *Newman's Lemma.* See [Bar84], p. 58.

Finally we add up β^2, which forms a confluent system that commutes with $\overset{\beta\eta^2\pi*}{\longrightarrow}$. So at last we can use Hindley-Rosen's Lemma[11], and we get confluence for the full system $\overset{\beta^2\eta^2\pi*}{\longrightarrow}$.

2.3.3 A general criterion for confluence

To get the confluence of $\overset{\beta\eta^2\pi*}{\longrightarrow}$ from the confluence of its restriction to *gentop* normal forms, we apply the following general method. Recall that two reduction systems R and S are said to commute when, for every term P, if $P \overset{R}{\twoheadrightarrow}{}^* Q$ and $P \overset{S}{\twoheadrightarrow}{}^* Q'$, there exists a term Q" such that $Q \overset{S}{\twoheadrightarrow}{}^* Q"$ and $Q' \overset{R}{\twoheadrightarrow}{}^* Q"$.

Lemma 2.3.5 *Let R be a reduction system that can be split in two subsystems R1 and R2 such that*

1. *R1 is weakly normalizing,*

2. *the set of R1 normal forms is closed w.r.t R2 reductions,*

3. *R2 is confluent on R1 normal forms, and*

4. $\overset{R}{\twoheadrightarrow}{}^*$ *commutes with* $\twoheadrightarrow R1$ *(see Notation 2.1.1).*

Then R is confluent.

Proof. Under the hypothesis above, any two reductions $\overset{R}{\twoheadrightarrow}{}^*$ starting from the same term can be completed to the commuting diagram shown in Figure 2.4:

- (1) ensures the existence of the R1 normal forms; hence we can build the central vertical arrow in the diagram (R1*| denotes reduction to some R1 n.f.).

- (4) ensures the existence and commutation of the upper inner rhombuses.

- (2) shows that the lower diagonal arrows in the upper rhombuses are made up of R2 reductions on R1 n.f.'s only, so that (3) guarantees the commutation of the lower inner rhombus.

Finally, the commutativity of the outermost rhombus follows from the commutativity of the inner rhombuses. □

---

[11] Hindley-Rosen's Lemma asserts the obvious but useful property that two separately confluent, commuting subsystems form a confluent system.

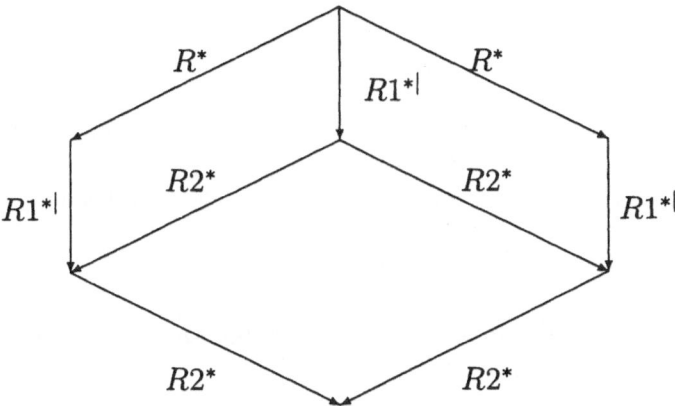

Figure 2.4: The factorization of confluence.

Val Breazu-Tannen pointed out to us that he used a particular case of this very same technique in [BT88], to prove Theorem 2.3, even if it was not singled out as a general tool for confluence like here. Later, Val Breazu-Tannen and Jean Gallier generalized in [BTG94] this Theorem to polymorphic lambda calculus, and there too Theorem 4.3 is clearly a particular instance of this technique. In our opinion, this independent discovery and use of this simple tool stresses its usefulness.

Notice that in the third condition, if the second condition holds, one may replace R2 by R. More interestingly, if we know that R1 is SN and confluent, then writing $R1(M)$ for the R1 normal form, we get the following alternative formulation of (4):

(4') If $M \xrightarrow{R} M'$, then $R1(M) \xrightarrow{R}{}^* R1(M')$

where the second occurrence or R might as well be R2, and the first occurrence of R might as well be R*. This form of the general criterion was already discovered and used by T. Hardin in her investigations of confluence properties of categorical combinators [Har89].

Our travel is close to the end. We shall take $\xrightarrow{\beta \eta^2 \pi *}{}^*$ as R, *gentop* as R1, and $\xrightarrow{\beta \eta^2 \pi *}$ less *gentop* as R2 and prove the four conditions of the criterion. The confluence of R2 on R1 normal forms is proved by establishing WCR and SN.

2.4 Confluence

Let in the following R stand for one of $\overset{\beta\eta^2\pi*}{\longrightarrow}$ $\overset{\beta\eta^2*}{\longrightarrow}$ $\overset{\beta\eta\pi*}{\longrightarrow}$ or $\overset{\beta\eta*}{\longrightarrow}$, R1 be
gentop and R2 be R less *gentop*. It will be intended that in the case of $\overset{\beta\eta\pi*}{\longrightarrow}$
and $\overset{\beta\eta*}{\longrightarrow}$, we consider only first-order terms and types and hence only the
corresponding restricted form of *gentop*, for which the following proofs hold
almost unchanged.

We first introduce some notation.

Notation 2.4.1 *We will note* $(M)^T$ *the gentop n.f. of a term* M *and* $\overset{gentop}{\underset{*}{\rightarrow}}$
the reduction to gentop normal form $(M)^T$.

Lemma 2.4.2 *The following equalities hold:*

1. $(PQ)^T = (P)^T(Q)^T$ *if* $(PQ){:}A$ *and* $A \notin Iso(\mathbf{T})$

2. $(p_i P)^T = p_i(P)^T$ *if* $p_i P{:}A$ *and* $A \notin Iso(\mathbf{T})$

3. $(\lambda x.P)^T = \lambda x.(P)^T$

4. $(\langle P, Q\rangle)^T = \langle (P)^T, (Q)^T\rangle$

5. $(\lambda X.P)^T = \lambda X.(P)^T$

6. $(P[B])^T = (P)^T[B]$ *if* $P[B]{:}A \notin Iso(\mathbf{T})$.

Proof. We only check 3 and leave the rest to the reader. Let $\lambda x.P :$
$A \to B$. If $A \to B\notin Iso(\mathbf{T})$, then the result is trivial, otherwise $(\lambda x.P)^T =$
$rep(A \to B) = \lambda z.(P)^T$ for some fresh variable z. Since the *gentop* normal
form of any term has no occurrence of variables in it (easily shown by
induction), then $\lambda z.(P)^T$ is equal to $\lambda x.(P)^T$ by α-conversion. $\square$

Lemma 2.4.3 $\overset{gentop}{\underset{*}{\rightarrow}}$ *is compatible with substitution, i.e.,*

$$(M[N/x])^T = (M)^T[(N)^T/x]$$

Proof. By an easy induction on the structure of M (see Table 2.1).
Notice that the case $M : U$ and $U \in Iso(\mathbf{T})$ is trivial since in both cases
the normal form is $rep(U)$, so in the table we consider only the case when
the normal form of a compound term is the combination of the normal
forms of its components.

$\square$

Lemma 2.4.4 *If $M \xrightarrow{R} M'$ then $(M)^T \xrightarrow{R}_= (M')^T$.*

Proof. We will proceed by induction on the structure of M. Notice that whenever M is a *gentop* redex, the claim holds trivially since the reductions we consider all preserve the type of the redex; so the type of M' is the same as that of M and their *gentop* normal forms are the same.[12] We shall thus assume that M is not a *gentop* redex. Furthermore, if the R reduction takes place in a proper subterm of M, the result follows easily by induction in each case (by Lemma 2.4.2), so we will not state it explicitly. We are left with the hypothesis that M is a redex which is not a *gentop* redex.

- M is a variable x. No reduction is possible, and the statement holds vacuously.

- M is an application. There is only one case:

    - M is $(\lambda x.P')Q$ and it β-reduces to $P'[Q/x]$.
      Then $(M)^T = ((\lambda x.P')^T(Q)^T) = (\lambda x.(P')^T)(Q)^T$, and it β-reduces to $(P')^T[(Q)^T/x]$, that is equal to $(P'[Q/x])^T$ by compatibility of $\overset{gentop}{\overset{*}{\to}}$ with substitution (Lemma 2.4.3).

- M is an abstraction. There are two cases:

    - M is $\lambda x.(Px)$ and it η-reduces to P. Then we have two possibilities for $(M)^T$ (notice that $(Px)^T = rep(V)$ is excluded as M would then be a *gentop* redex):

        * $\lambda x.((P)^T x)$ which η-reduces to $(P)^T$

        * $\lambda x.((P)^T rep(U))$ which η_{top}-reduces to $(P)^T$

    - M is $\lambda x.(Prep(U))$ and it η_{top}-reduces to P.
      Then $(M)^T = \lambda x.((P)^T rep(U))$ that η_{top}-reduces to $(P)^T$.

- M is a projection. The only case to consider is

---

[12]Remember that the contractum of a *gentop* redex depends only on the type of the redex, not on its structure.

  – M is $p_i \langle P_1, P_2 \rangle$ and it π-reduces to P_i.
  Then $(M)^T$ is $p_i(\langle P_1, P_2 \rangle)^T$, that is, $p_i \langle (P_1)^T, (P_2)^T \rangle$, which π-reduces to $(P_i)^T$.

- M is a pair. There are three cases:

  – M is $\langle p_1 P, p_2 P \rangle$ and it SP-reduces to P. By the previous lemma, we focus only on the following three possibilities for $(M)^T$:

    * $\langle p_1(P)^T, p_2(P)^T \rangle$ that SP-reduces to $(P)^T$

    * $\langle p_1(P)^T, rep(V) \rangle$ that SP_{top}-reduces to $(P)^T$

    * $\langle rep(U), p_2(P)^T \rangle$ that SP_{top}-reduces to $(P)^T$

  – M is $\langle p_1 P, rep(V) \rangle$ and it SP_{top}-reduces to P.
  Then $(M)^T$ is $\langle p_1(P)^T, rep(V) \rangle$ that SP_{top}-reduces to $(P)^T$

  – M is $\langle rep(U), p_2 P \rangle$ and it SP_{top}-reduces to P.
  Then $(M)^T$ is $\langle rep(U), p_2(P)^T \rangle$ that SP_{top}-reduces to $(P)^T$.

- M is an abstraction $\lambda t.P$. There is only one case to consider, namely P is $P'[X]$ and reduces to P' via η^2. We can assume $P'[X]$ not to be a *gentop* redex, as otherwise $M = \lambda X.P'[X]$ would be a *gentop* redex too, while we already factored out the case $M{:}U \in Iso(\mathbf{T})$. By Lemma 2.4.2, $(M)^T = (\lambda X.P'[X])^T = \lambda X.(P'[X])^T = \lambda X.(P')^T[X]$, that reduces via η^2 to $(P')^T$, as required.

Hence we have shown that $(M)^T \xrightarrow{R}_= (M')^T$. $\square$

Using the criterion for confluence, we will now show

Theorem 2.4.5 R *is confluent.*

Proof. We check the four hypotheses of Lemma 2.3.5 for R split in R1 and R2 as above.

1. gentop *is a strongly normalizing confluent reduction system.*

Proof. Each *gentop* step strictly decreases the number of *gentop* re-dexes in the term it is applied to. Since it is also trivially WCR, Newman's Lemma applies and we get CR too. □

2. *R2 reductions do not create new* gentop *redexes.*

Proof. By cases on the rule which is used. For all rules but β the result obviously follows from the fact that the reduct is a subterm of the redex. The case β is settled by noticing that, if M and N are in *gentop* n.f., then $M[N/x]$ is in *gentop* n.f. too. Indeed, this last property can be easily shown by induction on the structure of M.

If M is x or if it does not contain x free, then $M[N/x]$ is either M or N and the result follows from the hypothesis. We can also rule out the case where M is $rep(A)$, as then it has no free variables. So $M:A \notin Iso(\mathbf{T})$. If $M[N/x]$ contains a *gentop* redex P, then P cannot be $M[N/x]$, which has the same type as M, so P must be a proper subterm of $M[N/x]$. P cannot be a subterm of N either, or an unchanged subterm of M, since they are already in normal form, so it must be $M'[N/x]$ with M' a proper subterm of M containing a free occurrence of x. But M' is in *gentop* normal form as M is, hence, by induction hypothesis $M'[N/x]$ is not a *gentop* redex, so $M[N/x]$ is in *gentop* n.f. □

3. *The systems* $\xrightarrow{\beta\eta^2\pi*}$, $\xrightarrow{\beta\eta^2*}$, $\xrightarrow{\beta\eta\pi*}$ *and* $\xrightarrow{\beta\eta*}$ *are confluent over* gentop *normal forms.*

Proof. All the systems introduced so far are weakly confluent. We will prove in the appendix (Theorem 3.1.19, which follows closely the proof plan of [GLT90]), that $\xrightarrow{\beta\eta^2\pi*}$ is strongly normalizing over *gentop* normal forms. This implies strong normalization (over *gentop* normal forms) for all the others subsystems of it. Hence they are confluent over *gentop* n.f.'s by Newman's Lemma. □

Corollary 2.4.6 *R2 is confluent on* gentop *n.f.'s*

Proof. The previous point tells us that if $M \xrightarrow{R}* M'$ and $M \xrightarrow{R}* M''$, where M is in *gentop* normal form, then we can find M''' s.t. $M' \xrightarrow{R}* M'''$ and $M'' \xrightarrow{R}* M'''$. Now the second point shows that any reduction path starting from a *gentop* n.f. cannot contain *gentop* reductions, so the R reductions are made up only of R2 steps and we get the result. □

4. If $M \xrightarrow{R}{}^* M'$ then for any gentop n.f. N of M and N' of M', then we have that $N \xrightarrow{R}{}^* N'$.

Proof. By Lemma 2.4.4 above and a simple diagram chase. $\square$

Remark 2.4.7 *Again, this statement holds for all the reduction systems we are considering, as we showed it for* $\xrightarrow{\beta\eta^2\pi*}$, *and the statements for the others ones are particular cases of it.*

We can finally conclude, by Lemma 2.3.5, that R is confluent. $\square$

We still have a gap to fill for the second-order systems, since we have left out β^2. We shall prove CR for $\xrightarrow{\beta^2\eta^2\pi*}$ and $\xrightarrow{\beta^2\eta^2*}$ by using Hindley-Rosen's Lemma.

Let R1 be the system $\xrightarrow{\beta\eta^2\pi*}$ (or $\xrightarrow{\beta\eta^2*}$) and R2 be β^2.

Lemma 2.4.8 β^2 *is confluent.*

Proof. The system consisting of β^2 alone satisfies the diamond property, i.e., every one-step divergence can be closed in one step, hence is CR. $\square$

We just proved that R1 is CR (Theorem 2.4.5), so we are left to show that R1 commutes with R2, and the CR property will follow by Hindley-Rosen's Lemma.

Theorem 2.4.9 *R1 and R2 commute with each other.*

Proof. It suffices to prove that, if $M \xrightarrow{R1} M'$ and $M \xrightarrow{R2} N$, then there exists a term M'' s.t. $N \xrightarrow{R2}{}^* M''$ and $M' \xrightarrow{R1}{}_= M''$ (see Lemma 3.3.6 in [Bar84], p. 65). The only superpositions arise with η^2 and *gentop*, and are easily closed up, so that it suffices to notice that β^2 cannot duplicate existing redexes (β^2 can only duplicate types, that are not redexes), so that the constraint on the R1 reduction that closes the diagram gives no problem. The details are left to the reader. $\square$

So we finally get, by Hindley-Rosen's Lemma.

Theorem 2.4.10 *The systems* $\xrightarrow{\beta^2\eta^2\pi*}$ *and* $\xrightarrow{\beta^2\eta^2*}$ *are confluent* [13].

---

[13]We also found an alternative proof of the confluence of $\xrightarrow{\beta^2\eta^2*}$ that does not extend to the case with SP. It relies on yet another splitting of the rules, taking *gentop* and the β rules on one hand, and the *eta-like* rules on the other. The proof uses the same criterion for confluence as we used in this section. In order to check the last condition, we rely on a parallelization of R2, which does not work well when the nonlinear surjective pairing rule is added to R2 (cf. introduction). So we abandoned that proof technique which we were not able to extend to the full system.

2.5 Weak normalization

For the first-order systems, we get from the previous section a normalizing strategy for free: first go to the *gentop* normal form, then use the SN property on *gentop* normal forms.

Summarizing, we have obtained:

Theorem 2.5.1 *The calculi $\lambda^1 \beta \eta *$, $\lambda^1 \beta \eta \pi *$ are effectively weakly normalizing.*

Since for the second-order systems we have left out β^2 and η^2, we find them on the way; we can deal with them at the price of a splitting of the set of rules that is different from the splitting which lead us to confluence.

Theorem 2.5.2 *The calculi $\lambda^2 \beta \eta *$, $\lambda^2 \beta \eta \pi *$ are effectively weakly normalizing.*

Proof. The reduction system R can be split into the two subsystems $R1 = \{\beta, \pi, gentop, \beta^2, \eta^2\}$ and $R2 = \{\eta, SP, \eta_{top}, SP_{top}\}$. R1 is shown to be SN by a straightforward adaptation of the technique of [GLT90] (see Section 3.2).

One can show by an easy induction on the structure of the context surrounding an R2 redex that no R2 reduction creates any *new* R1 redex.

Theorem 2.5.3 *R2 reductions do not create new R1 redexes.*

Proof. It suffices to consider the case of $\lambda^2 \beta \eta \pi *$ as the R1 and R2 systems for it embodies the R1 and R2 systems for all the others.

First notice that since reductions preserve the type, no new *gentop* redex can be created since *gentop* redexes depend only on the type of the terms. As for β, π, β^2 and η^2, let $P \xrightarrow{R2} P'$.

A context with a single hole for our calculus can be defined inductively as follows:

$$C[] := [] \mid (QC[]) \mid (C[]Q) \mid p_i C[] \mid \lambda x.C[] \mid \langle Q, C[] \rangle \mid \langle C[], Q \rangle \mid \lambda X.C[] \mid C[][A]$$

We prove the lemma by induction on the context $C[]$ where the R2 redex P occurs. Notice that the only interesting cases are when P appears in a position where a new R1 redex can be created, i.e., when it is applied to a term or it appears in $p_i P$.

- $[]$ trivial since P' is a subterm of P for all rules in R2

- $(QC[])$ by induction hypothesis, $C[P']$ contains no R1 redexes not appearing in $C[P]$. Since the fact that the application $(QC[])$ is a redex which depends only on Q, which does not change, and redexes inside Q do not change too, we are done.

- $(C[]Q)$ by induction hypothesis, $C[P']$ contains no R1 redexes not appearing in $C[P]$. Q does not change, so redexes inside Q also do not change. The only possible new redex would be the application $(C[P']Q)$ if $C[P']$ is an abstraction and $C[P]$ is not. This can happen only if $C[P]$ is P, and for typing reasons, it means $(PQ) \overset{\eta}{\longrightarrow} (P'Q)$ or $(PQ) \overset{\eta_{top}}{\longrightarrow} (P'Q)$. In both cases P is already an abstraction, so this redex is not new either and we are done.

- $p_i C[]$ by induction hypothesis, $C[P']$ contains no R1 redexes not appearing in $C[P]$. The only possible new redex would be $p_i C[P']$ if $C[P']$ is a pair and $C[P]$ is not. Again, this can happen only if $C[P]$ is P, and due to typing reasons, it means $P \overset{SP}{\longrightarrow} P'$ or $P \overset{SP_{top}}{\longrightarrow} P'$. In both cases P is already a pair, so this redex is not new either and we are done.

- $\lambda x.C[]$ by induction hypothesis, $C[P']$ contains no R1 redexes not appearing in $C[P]$. Since an abstraction is not an R1 redex, the same holds for $\lambda x.C[P]$.

- $\langle Q, C[]\rangle$, $\langle C[], Q\rangle$, $\lambda X.C[]$, $C[][A]$: similarly as for abstraction.

$\square$

This has the following important consequence:

Corollary 2.5.4 *The set of R1 normal form is closed w.r.t. R2 reductions.*

Since R2 is obviously SN, as the rules strictly decrease the size of the terms they apply to, this corollary gives us the following, very easy, effective normalizing (standard) strategy. Given a term M,

1. first R1–normalize it reaching, say, M',

2. then R2–normalize M' reaching, say, M''.

M'' is the desired normal form. $\square$

The previous result about weak normalization for the first-order fragment can obviously be derived as a corollary from this theorem, but we actually needed the ingredients of the previous proof to get the confluence of our systems.

2.6 Decidability and conservative extension results

From the confluence and weak normalization for our calculi, it is now easy to get also the decidability of the associated equational theories as well as conservativity results.

Corollary 2.6.1 *The equational theories for* $\lambda^1\beta\eta*$, $\lambda^1\beta\eta\pi*$, $\lambda^2\beta\eta*$ *and* $\lambda^2\beta\eta\pi*$ *are decidable.*

Proof. Given terms M and N, consider their normal forms M' and N' (they exist by WN). If $M = N$, then (by CR) M' is syntactically equal to N'. So, to decide equality it suffices to take the normal forms (that is, effective, as we provided a normalizing strategy for each one of these calculi) and to check if they are equal. $\square$

Corollary 2.6.2 *(Conservative extensions)* *For* L *any of the calculi* $\lambda^2\beta\eta\pi*$ $\lambda^2\beta\eta*$ $\lambda^1\beta\eta\pi*$ *or* $\lambda^1\beta\eta*$ *call* $\xrightarrow{L}$ *the rewriting system corresponding to* L, *that is,* $\xrightarrow{\beta^2\eta^2\pi*}$, $\xrightarrow{\beta^2\eta^2*}$, $\xrightarrow{\beta\eta\pi*}$ *or* $\xrightarrow{\beta\eta*}$. *Let* L' *be a subtheory of* L *which has the following stability property. If* M *is in the sublanguage of* L' *and* $M \xrightarrow{L} N$, *then* N *is also in* L', *and* M *and* N *are provably equal in* L'. *If* M *and* N *are terms of* L' *that are equal in* L, *then they are already equal in* L'.

Proof. If M and N are equal in L, then, by the CR property there exists a term P s.t. M and N reduce to P in L. But M and N are terms of L', and no reduction in any of the calculi we consider can reach terms outside L'. Then the reductions $M \xrightarrow{L}^* P$ and $N \xrightarrow{L}^* P$ correspond to provable equations in L', so that M is equal to N in L'. $\square$

In [BDCL92] (here in Chapter 4), for example, we need the conservativity of the equational theory of $\lambda^1\beta\eta\pi*$ over the simple typed λ-calculus, while in [DC94] (here in Chapter 4), we actually use the conservativity of $\lambda^2\beta\eta\pi*$ over the second-order lambda calculus.

2.7 Other related works

As far as we know, our results are new for what concerns polymorphism, while other proofs of Corollary 2.6.1 have been given in the literature, for the case of the first-order calculi. We already gave a broad overview of the proofs based on confluence and normalization in the introduction, but there have been several interesting related works that managed to prove

decidability (and in some cases conservativity, too) without giving an explicit confluent and normalizing rewriting system for the full calculi. The
method used in [LS86], for example, is based on

- the elimination of **T**

- a proof of confluence via WCR and SN (WCR holds there without a
  need to add funny rules, and the computability method works well
  without special restrictions, as was first shown by R. De Vrijer).

Another method, which was found independently by A.S. Troelstra (see
[Tro86], where it is used to prove SN rather than CR) and T. Hardin (see
[Har89]), goes further by eliminating products as well as **T**. The two methods allow to prove conservativity as well as decidability, but the overall
construction is quite tedious. Let us be more specific, since the explana-
tions provided by Lambek and Scott, in [LS86] pp. 81–82, are somewhat
handwaving. The exploitation of the type isomorphisms can be formalized
as follows. To every type T we associate a **T**-*free* type T°.

Definition 2.7.1 *For any type T, we define its "top-free" form T° as the
normal form of T w.r.t. the following (confluent and strongly normalizing)
type rewrite system $\leadsto$:*

$$A \times \mathbf{T} \leadsto A \qquad\qquad \mathbf{T} \times A \leadsto A$$
$$\mathbf{T} \to A \leadsto A \qquad\qquad A \to \mathbf{T} \leadsto \mathbf{T}$$

Thus a "**T**-free" type is either **T**, or a type where **T** does not occur. Then
one may extend this mapping to terms, so that for a term $M{:}A$ we have
$M^\circ{:}A^\circ$, in such a way that

$$M =_{\beta\eta\pi*} N \Longleftrightarrow M^\circ =_{\beta\eta\pi} N^\circ.$$

Similarly, to a type A of $\lambda^1\beta\eta\pi*$ we can associate a sequence of types A^*
constructed from type variables with the arrow only, and to a term M a
sequence M^* of terms of the types that appear in A^*.

Then $M =_{\beta\eta\pi*} N$ iff $M_1 =_{\beta\eta} N_1, \cdots, M_n =_{\beta\eta} N_n$, where $M^* = M_1, \cdots, M_n$
and $N^* = N_1, \cdots, N_n$.

This formalizes the assertion of Lambek and Scott that there is "no loss
of generality", as far as decision is concerned, if one removes the terminal
object (or both the terminal object and the products).

Moreover, these translations of types and terms are conservative in the
sense that if A is a type where **T** (respectively, **T** and $\times$) does not occur
and $M{:}A$, then A° and M° (respectively A^* and M^*) are just A and M.
Corollary 2.6.2 is an immediate consequence of this.

Yet another solution to the decidability problem for equational theories of cartesian closed categories has been proposed by A. Obtulowicz [Obt87]. His approach is very algebraic in nature. Obtulowicz effectively defines operations on some canonical forms, turning the set of canonical forms into an initial algebra. Then, to decide that two terms are equal, one computes their interpretation in the initial algebra and checks whether the resulting canonical forms coincide. This approach is very technical and contains hidden rewriting techniques. But it is interesting, because it does not *a priori* require such strong assumptions as to find a noetherian and confluent rewriting system.

Anyway, A. Obtulowicz did not show decidability for exactly the same equational theories as we do here. Specifically, he deals with the critical pairs which lead us to the SP_{top} rules in a different way. He forces an equational theory on types as well as on terms. Specifically, the canonical type isomorphisms underlying the translation ° above are forced to be true equalities (and models of these theories have thus to identify on the nose, say $A \times \mathbf{T}$ and A). New equations between terms are added, which witness these identifications at the level of terms. Here is one of them:

$$\langle M, * \rangle = M \text{ for } M : A \times \mathbf{T}$$

With the aid of this equation and of one of its consequences, namely

$$\mathrm{p}_1 M = M \text{ for } M : A \times \mathbf{T} ,$$

one can solve the critical pair

$$\langle \mathrm{p}_1 M, * \rangle \leftarrow \langle \mathrm{p}_1 M, \mathrm{p}_2 M \rangle \rightarrow M$$

by just noting that $\langle \mathrm{p}_1 M, * \rangle \rightarrow \mathrm{p}_1 M \rightarrow M$. It would be worthwhile to investigate these theories from a rewriting point of view.

Another treatment of the terminal object with identification of types can be found in [Nip90], which is only concerned with *local* confluence and does not take into account the commutativity of product. His work seems extremely interesting in a type inference perspective, but is less related to decidability of the equality problem we deal with here.

M	LHS	RHS
x	$(N)^T$	$(N)^T$
y	y	y
(PQ)	$((PQ)[N/x])^T$	$(PQ)^T[(N)^T/x]$
	$= (P[N/x]Q[N/x])^T$	$= ((P)^T(Q)^T)[(N)^T/x]$
	$= ((P[N/x])^T(Q[N/x])^T)$	$= ((P)^T[(N)^T/x](Q)^T[(N)^T/x])$
	$= ((P)^T[(N)^T/x])((Q)^T[(N)^T/x])$	
$\lambda y.P$	$(\lambda y.P[N/x])^T$	$(\lambda y.P)^T[(N)^T/x]$
	$= \lambda y.(P[N/x])^T$	$= (\lambda y.(P)^T)[(N)^T/x]$
	$= \lambda y.(P)^T[(N)^T/x]$	$= \lambda y.(P)^T[(N)^T/x]$
$p_i P$	$(p_i P[N/x])^T$	$(p_i P)^T[(N)^T/x]$
	$= p_i(P[N/x])^T$	$= p_i(P)^T[(N)^T/x]$
	$= p_i(P)^T[(N)^T/x]$	$= p_i(P)^T[(N)^T/x]$
$\langle P,Q \rangle$	$(\langle P[N/x], Q[N/x] \rangle)^T$	$(\langle P,Q \rangle)^T[(N)^T/x]$
	$= \langle (P[N/x])^T, (Q[N/x])^T \rangle$	$= \langle (P)^T, (Q)^T \rangle[(N)^T/x]$
	$= \langle (P)^T[(N)^T/x], (Q)^T[(N)^T/x] \rangle$	$= \langle (P)^T[(N)^T/x], (Q)^T[(N)^T/x] \rangle$
$\lambda t.P$	$(\lambda t.P[N/x])^T$	$(\lambda t.P)^T[(N)^T/x]$
	$= \lambda t.(P[N/x])^T$	$= (\lambda t.(P)^T)[(N)^T/x]$
	$= \lambda t.(P)^T[(N)^T/x]$	$= \lambda t.(P)^T[(N)^T/x]$
$P[A]$	$(P[A][N/x])^T$	$(P[A])^T[(N)^T/x]$
	$= (P[N/x][A])^T$	$= (P)^T[A][(N)^T/x]$
	$= (P[N/x])^T[A]$	$= (P)^T[(N)^T/x][A]$
	$= (P)^T[(N)^T/x][A]$	

Table 2.1: Compatibility of *gentop* n.f. with substitution.

Chapter 3

Strong normalization for subsystems of $\lambda^2\beta\eta\pi*$

Our proof of confluence in Theorem 2.4.5 relies upon the strong normalization of $\overset{\beta\eta^2\pi*}{\longrightarrow}{}^*$ over the set of *gentop* normal forms, while we need the strong normalization of $\overset{\beta^2\eta^2\pi*}{\longrightarrow}{}^*$ less η_{top} and SP_{top} over the full set of terms in order to provide an effective weakly normalizing strategy for $\overset{\beta^2\eta^2\pi*}{\longrightarrow}{}^*$ in Theorem 2.5.2.

This chapter provides these two proofs of strong normalization in Sections 3.1 and 3.2 respectively, by suitably adapting one of the various versions of the reducibility method. We choose here to apply Girard's method, following essentially the same proof plan as in [GLT90], p. 42-47. Since there is almost no difference in the proofs for the two systems, we will detail the first one only and point out the differences for the second case.

As pointed out in Section 2.3.2, the reducibility method fails for the full system where η_{top} and SP_{top} are allowed to freely interact with any term of the calculus; we are not able to deal in the crucial proofs of the abstraction and pairing lemmas (Lemmas 3.1.13 and 3.1.12) with some reductions that arise in the full system.

To rule out these reductions, one can either restrict the system to *gentop* normal forms only (this requires in turn to rule out the β^2 rule, which does not preserve *gentop* normal forms, as shown in Example 2.3.4), or one can simply rule out η_{top} and SP_{top}.

3.1   Normalization without β^2 on *gentop* n.f.'s

In this section we will show that the system $\overset{\beta\eta^2\pi*}{\longrightarrow}{}^*$ (the full system $\overset{\beta^2\eta^2\pi*}{\longrightarrow}{}^*$ less β^2) is strongly normalizing over the set of *gentop* normal forms. This means that all along the proof *any* gentop *reduction is ruled out*, so we will not explicitly state all the time that *gentop* reductions cannot occur. Moreover, to improve readability, $\longrightarrow$ will stand for $\overset{\beta\eta^2\pi*}{\longrightarrow}{}^*$ in this section.

Definitions

Definition 3.1.1 (neutral terms) *A term $t{:}U$ is neutral iff one of the following conditions is satisfied:*

- *if $U \notin Iso(\mathbf{T})$ and t is not an abstraction, a type abstraction or a pair, or*

- *if $U \in Iso(\mathbf{T})$ (then t is rep(U), as we consider only terms in* gentop *normal form).*

Definition 3.1.2 (longest reduction path for a term) *For a term u, $\nu(u)$ denotes the length of the longest reduction path starting from u. Notice that, by König's Lemma, if u is strongly normalizable, then $\nu(u)$ is finite.*

Definition 3.1.3 *A reducibility candidate of type U is a set R of terms of type U with the following properties.*

CR1 if $t \in R$, then t is strongly normalizable.

CR2 if $t \in R$ and $t \to t'$, then $t' \in R$.

CR3 if t is neutral and for all t' s.t. $t \to t'$ we have that $t' \in R$, then $t \in R$.

Remark 3.1.4 *A reducibility candidate R of type U is never empty:*

- *If $U \in Iso(\mathbf{T})$, then rep(U) is neutral and in normal form and hence belongs to R by (CR3).*

- *If $U \notin Iso(\mathbf{T})$, then any variable of type U is neutral and in normal form and hence belongs to R by (CR3).*

Proposition 3.1.5 *The set of strongly normalizable terms of type U is a reducibility candidate.*

Proof.

- (CR1) is a tautology.

- (CR2) if t is strongly normalizable, then every t' s.t. $t \longrightarrow t'$ is strongly normalizable.

- (CR3) every reduction path leaving t must pass through one of the terms t' that are one step from t. Since all t' are strongly normalizable, then t is strongly normalizable also.

□

Definition 3.1.6 (product and arrow of reducibility candidates) *If R and S are reducibility candidates of types U and V, we can define sets $R \to S$ of terms of type $U \to V$ and $R \times S$ of terms of type $U \times V$ as follows:*

$\quad$ – for all $u \in R$, $(tu) \in S$ if $V \notin Iso(\mathbf{T})$

$\quad$ – $t = rep(U \to V)$ if $V \in Iso(\mathbf{T})$

- $t \in R \times S$ (of type $U \times V$) $\Longleftrightarrow$

$\quad$ – $p_1 t \in R$ and $p_2 t \in S$ if U, V are not in $Iso(\mathbf{T})$

$\quad$ – $p_1 t \in R$ if $U \notin Iso(\mathbf{T})$, $V \in Iso(\mathbf{T})$

$\quad$ – $p_2 t \in S$ if $U \in Iso(\mathbf{T})$, $V \notin Iso(\mathbf{T})$

$\quad$ – $t = rep(U \times V)$ if U, $V \in Iso(\mathbf{T})$

Remark 3.1.7 *Notice that, as t and u are in* gentop *normal form, and due to the conditions on U and V, the terms (tu), $p_1 t$ and $p_2 t$ above are still in* gentop *normal form.*

Theorem 3.1.8 *If R_1 and R_2 are reducibility candidates of types U_1 and U_2, then $R_1 \times R_2$ and $R_1 \to R_2$ are reducibility candidates of type $U_1 \times U_2$ and $U_1 \to U_2$ respectively.*

Proof. Assume that R_1 and R_2 are reducibility candidates of type U_1 and U_2, respectively.

1. $R_1 \times R_2$ is a reducibility candidate of type $U_1 \times U_2$. If $U_1 \times U_2 \in Iso(\mathbf{T})$, then (CR1), (CR2) and (CR3) hold vacuously due to the fact that we consider only *gentop* normal forms, so let's assume in the following that $U_1 \notin Iso(\mathbf{T})$ and/or $U_2 \notin Iso(\mathbf{T})$.

   - (CR1) if $t \in U_1 \times U_2$ and $U_i \notin Iso(\mathbf{T})$, then $p_i t$ is strongly normalizable by the induction hypothesis on U_i, since $p_i t \in U_i$ by definition. Hence t is strongly normalizable.

   - (CR2) if $t \longrightarrow t'$, then $p_1 t \longrightarrow p_1 t'$ and/or $p_2 t \longrightarrow p_2 t'$. As $t \in U_1 \times U_2$, then $p_1 t \in U_1$ and/or $p_2 t \in U_2$. By induction hypothesis **CR2** for U_1 and/or U_2 we get $p_1 t' \in U_1$ and/or $p_2 t' \in U_2$ and hence, by definition, $t' \in U_1 \times U_2$.

   - (CR3) t is neutral and all t' one step from t are in $U_1 \times U_2$. We need to show $p_1 t \in U_1$ and/or $p_2 t \in U_2$. Now notice that applying a conversion inside $p_i t$ can only result in some $p_i t'$ as t is not a pair (it is neutral and it is not $rep(U_1 \times U_2)$). But $p_1 t' \in U_1$ and/or $p_2 t' \in U_2$ as t' is in $U_1 \times U_2$. In any case, $p_1 t$ and/or $p_2 t$ are neutral and every term one step from it is in $U_1 \times U_2$, so the induction hypothesis for U_1 and/or U_2 ensure $p_1 t \in U_1$ and/or $p_2 t \in U_2$. So $t \in U_1 \times U_2$.

2. $R_1 \to R_2$ is a reducibility candidate of type $U_1 \to U_2$.

We can assume that $U_2 \notin Iso(\mathbf{T})$ since otherwise $U_1 \to U_2 \in Iso(\mathbf{T})$, and then (CR1), (CR2) and (CR3) hold vacuously.

- (CR1) if $t \in U_1 \to U_2$, then let u be a variable x of type U_1 if $U_1 \notin Iso(\mathbf{T})$ or else $rep(U_1)$. Since $u \in any$ reducibility candidate, (remark 3.1.4), we get that $(tu) \in U_2$ by definition, hence (tu) is strongly normalizable by induction hypothesis for U_2, that suffices to show that t is strongly normalizable.

- (CR2) if $t \longrightarrow t'$, we need to show $(t'u) \in U_2$ for all $u \in U_1$. Take then $u \in U_1$; we have $(tu) \in U_2$ and $(tu) \longrightarrow (t'u)$, and hence $(t'u) \in U_2$ by induction hypothesis on U_2.

- (CR3) t is neutral and all t' one step from t are in $R_1 \to R_2$. In order to show $t \in U_1 \to U_2$, we need to show $(tu) \in U_2$ for all $u \in U_1$. By induction hypothesis on U_1, we get u is strongly normalizable, so we can argue by induction on $\nu(u)$. In one step, (tu) converts to:

  - $(t'u)$ with t' one step from t.
    As $t' \in U_1 \to U_2$, we get $(t'u) \in U_2$ by definition.
  - (tu') with u' one step from u.
    By induction hypothesis on U_1, $u' \in U_1$ and $\nu(u') < \nu(u)$, so $(tu') \in U_2$ by the induction hypothesis on u.
  - there is no other possibility, as t is already in *gentop* n.f. and it is neutral, hence not of the form $\lambda x.v$ (it cannot be $rep(U_1 \to U_2)$ as we already assumed $U_1 \to U_2 \notin Iso(\mathbf{T})$).

□

3.1.1   Reducibility with parameters

Let T be a type, and let $\overrightarrow{X}$ be a set of type variables containing at least all the free type variables of T. For $\overrightarrow{U}$ a sequence of types of the same length, let $T[\overrightarrow{U}/\overrightarrow{X}]$ be the type obtained by simultaneous substitution of the X's with the U's, and let $\overrightarrow{R}$ a sequence of reducibility candidates of corresponding types.

Definition 3.1.9 *The set* $RED_T[\overrightarrow{R}/\overrightarrow{X}]$ *of reducible terms of type* $T[\overrightarrow{U}/\overrightarrow{X}]$ *is defined by induction on the type* T *as follows.*

- if T is atomic, $RED_T[\vec{R}/\vec{X}]$ is the set of strongly normalizable terms of type $T[\vec{U}/\vec{X}] = T$

- if T is X_i, $RED_T[\vec{R}/\vec{X}]$ is R_i

- if T is $U \times V$, then $RED_T[\vec{R}/\vec{X}]$ is $RED_U[\vec{R}/\vec{X}] \times RED_V[\vec{R}/\vec{X}]$

- if T is $U \to V$, then $RED_T[\vec{R}/\vec{X}]$ is $RED_U[\vec{R}/\vec{X}] \to RED_V[\vec{R}/\vec{X}]$

- if T is $\forall Y.W$, then $RED_T[\vec{R}/\vec{X}]$ is the set of terms t of type $[\vec{U}/\vec{X}]$ such that, for every type V and reducibility candidate S of this type, $t[V] \in RED_W[\vec{R}/\vec{U}, S/Y]$

Lemma 3.1.10 *rep(U) is normal for all $U \in Iso(\mathbf{T})$.*

Proof. By a straightforward induction on the structure of the term. $\square$

Theorem 3.1.11 $RED_T[\vec{R}/\vec{X}]$ *is a reducibility candidate of type $T[\vec{U}/\vec{X}]$*

Proof. We proceed by structural induction on the type T.

Since we consider only terms in *gentop* normal form, *there is no term of type U besides rep(U) if $U \in Iso(\mathbf{T})$*. Moreover, due to the previous lemma and the definition of reducibility, $rep(U)$ trivially satisfies (CR1), (CR2) and (CR3), so we will not consider explicitly the case of types in $Iso(\mathbf{T})$ in the induction.

Atomic types

If T is atomic, then $RED_T[\vec{R}/\vec{X}]$ is the set of strongly normalizing terms of type T, and we already proved it to be a reducibility candidate (Proposition 3.1.5).

Type Variables

If T is X_i, then $RED_T[\vec{R}/\vec{X}]$ is R_i, that is a reducibility candidate by definition.

Product types

Let T be $U_1 \times U_2$. Then $RED_T[\vec{R}/\vec{X}] = RED_{U_1}[\vec{R}/\vec{X}] \times RED_{U_1}[\vec{R}/\vec{X}]$ by definition. We can apply the induction hypothesis for $RED_{U_1}[\vec{R}/\vec{X}]$ and $RED_{U_2}[\vec{R}/\vec{X}]$, so that the result then follows by Theorem 3.1.8.

Arrow types

Let T be $U_1 \to U_2$. Then $RED_T[\vec{R}/\vec{X}] = RED_{U_1}[\vec{R}/\vec{X}] \to RED_{U_1}[\vec{R}/\vec{X}]$ by definition. We can apply the induction hypothesis for $RED_{U_1}[\vec{R}/\vec{X}]$ and $RED_{U_2}[\vec{R}/\vec{X}]$, so that the result then follows by Theorem 3.1.8.

Universal types

Let $T = \forall Y.W$. We can assume that $W \notin Iso(\mathbf{T})$ as otherwise $\forall Y.W \in Iso(\mathbf{T})$.

- (CR1) if $t \in RED_{\forall Y.W}[\vec{R}/\vec{X}]$, then let V be an arbitrary type and S be an arbitrary reducibility candidate of this type (for example, the strongly normalizable terms of type V). Then $t[V] \in RED_W[\vec{R}/\vec{X}, S/Y]$, and so, by induction hypothesis, we know that $t[V]$ is strongly normalizable. A fortiori t is strongly normalizable.

- (CR2) if $t \xrightarrow{\beta\eta\pi*} t'$, then for all types V and reducibility candidate S of this type, we have that $t[V] \in RED_W[\vec{R}/\vec{X}, S/Y]$ and $(t[V]) \xrightarrow{\beta\eta\pi*} (t'[V])$, hence $t'[V] \in RED_W[\vec{R}/\vec{X}, S/Y]$ by induction hypothesis on W. So, by definition, $t' \in RED_{\forall Y.W}[\vec{R}/\vec{X}]$.

- (CR3) t is neutral and all t' one step from t are in $RED_T[\vec{R}/\vec{X}]$. Take V and S: if we apply a conversion inside $t[V]$, the result is $t'[V]$ since t is neutral (and, again, not $rep(\forall Y.W)$, as $t \xrightarrow{\beta\eta\pi*} t'$). Now, $t'[V]$ is in $RED_W[\vec{R}/\vec{X}, S/Y]$ as t' is in $RED_T[\vec{R}/\vec{X}]$. By induction hypothesis, we get $t'[V] \in RED_W[\vec{R}/\vec{X}, S/Y]$, so $t \in RED_T[\vec{R}/\vec{X}]$.

□

Reducibility theorem

We shall need some lemmas to deduce reducibility of a term from reducibility of its subterms.

Lemma 3.1.12 *(Pairing) Let $u_1 \in RED_{U_1}[\vec{R}/\vec{X}]$ and $u_2 \in RED_{U_2}[\vec{R}/\vec{X}]$. Then $\langle u_1, u_2 \rangle \in RED_{U_1 \times U_2}[\vec{R}/\vec{X}]$.*

Proof. We can assume that $U_1 \notin Iso(\mathbf{T})$ and/or $U_2 \notin Iso(\mathbf{T})$, as otherwise $\langle u_1, u_2 \rangle = rep(U_1 \times U_2)$ and then $RED_{U_1 \times U_2}[\vec{R}/\vec{X}]$ is $\{rep(U_1 \times U_2)\}$.

We can argue by induction on $\nu(u_1) + \nu(u_2)$, by CR1, to show that, for $i = 1$ and/or $i = 2$, $p_i\langle u_1, u_2 \rangle \in RED_{U_i}[\vec{R}/\vec{X}]$.

Let $i = 1$ for simplicity. The term $p_1\langle u_1, u_2 \rangle$ converts to:

- u_1, which is in $RED_{U_1}[\vec{R}/\vec{X}]$ by hypothesis.

- $p_1\langle u', u_2 \rangle$ with u' one step from u_1.
  Then u' is in $RED_{U_1}[\vec{R}/\vec{X}]$ by CR2 and $\nu(u') < \nu(u_1)$, so $p_1\langle u', u_2 \rangle \in RED_{U_1}[\vec{R}/\vec{X}]$ by induction hypothesis.

- $p_1\langle u_1, v' \rangle$ with v' one step from u_2. We get $p_1\langle u_1, v' \rangle \in RED_{U_1}[\vec{R}/\vec{X}]$ as above.

- $p_1 w$ if u_1 is $p_1 w$ and u_2 is $p_2 w$.
  But $p_1 w = u_1$ is in $RED_{U_1}[\vec{R}/\vec{X}]$ by hypothesis.

- $p_1 w$ if u_1 is $p_1 w$ and u_2 is $rep(U_2)$.
  By definition of parametric reducibility for product types when one of the factor types is in $Iso(\mathbf{T})$, we get that $u_1 \in RED_{U_1}[\vec{R}/\vec{X}]$ as $p_1 w = u_1$ is in $RED_{U_1}[\vec{R}/\vec{X}]$ by hypothesis.

In every case, the neutral terms $p_i\langle u_1, u_2 \rangle$ convert to terms in $RED_{U_i}[\vec{R}/\vec{X}]$ only, for $i = 1$ and/or $i = 2$, so they are in $RED_{U_i}[\vec{R}/\vec{X}]$ by CR3. Hence $\langle u_1, u_2 \rangle$ is in $RED_{U_1 \times U_2}[\vec{R}/\vec{X}]$. □

Lemma 3.1.13 *(Abstraction) Let $x:U$ and $v:V$. If for all $u \in RED_U[\vec{R}/\vec{X}]$ we have that $v[u/x] \in RED_V[\vec{R}/\vec{X}]$, then $\lambda x.v \in RED_{U \to V}[\vec{R}/\vec{X}]$.*

Proof. We can assume that $V \notin Iso(\mathbf{T})$ as otherwise v is $rep(V)$, and $\lambda x.v$ is $rep(U \to V)$ as $U \to V \in Iso(\mathbf{T})$, and it is reducible by definition.

To show that $\lambda x.v \in RED_{U \to V}[\vec{R}/\vec{X}]$, we need to show that $(\lambda x.v)u \in RED_V[\vec{R}/\vec{X}]$ for all $u \in RED_U[\vec{R}/\vec{X}]$.

There are two cases: either $U \in Iso(\mathbf{T})$ or not.

In the first case, $v[u/x] = v$ as it is in *gentop* normal form, hence there is no free occurrence of x in v, and the only term u of type U is $rep(U)$. Since

$t = (\lambda x.v)u$ is neutral, it suffices to show that for every term t' one-step from it $t' \in RED_V[\vec{R}/\vec{X}]$. Since $v = v[rep(U)/x] \in RED_V[\vec{R}/\vec{X}]$ by hypothesis, hence strongly normalizing, we can argue by induction on $\nu(v)$. The one-step reducts of $(\lambda x.v)u$ are:

- $v[u/x]$ which is in $RED_V[\vec{R}/\vec{X}]$ by hypothesis

- $(\lambda x.v')u$ with v' one step from v. Then $v'[u/x]$ is in $RED_V[\vec{R}/\vec{X}]$ by CR2 as it is one step from $v[u/x]$ and we are done by induction hypothesis as $\nu(v') < \nu(v)$

- $(v'u)$ via η_{top} if $v = v'rep(U)$.
  Now, $u = rep(U)$ so $(v'u) = v'rep(U) = v = v[u/x]$ which is in $RED_V[\vec{R}/\vec{X}]$ by hypothesis.

In the second case, $x:U$ is in $RED_U[\vec{R}/\vec{X}]$ (Remark 3.1.4). So $v = v[x/x]$ is in $RED_V[\vec{R}/\vec{X}]$ and hence strongly normalizable by CR2, and we can argue by induction on $\nu(u) + \nu(v)$ to show that all terms one step from $(\lambda x.vu)$ are reducible. The one-step reducts of $(\lambda x.v)u$ are:

- $v[u/x]$ that is in $RED_V[\vec{R}/\vec{X}]$ by hypothesis.

- $(\lambda x.v')u$ with v' one step from v. Since $v'[u/x]$ is one step from $v[u/x]^1$, then it is in $RED_V[\vec{R}/\vec{X}]$ by CR2. Furthermore, $\nu(v') < \nu(v)$, so by induction hypothesis we get $(\lambda x.v'u) \in RED_V[\vec{R}/\vec{X}]$.

- $(\lambda x.v)u'$ with u' one step from u. Then $u' \in RED_U[\vec{R}/\vec{X}]$ by CR2, $\nu(u') < \nu(u)$ and $v[u'/x] \in RED_V[\vec{R}/\vec{X}]$ by repeated applications of CR2, as it is some step from $v[u/x]$. So we can apply again the induction hypothesis.

- $(v'u)$ via η if $\lambda x.v$ is $\lambda x.v'x$ and $x \notin FV(v')$.
  It is in $RED_V[\vec{R}/\vec{X}]$ as $v[u/x] = (v'u)$ is in $RED_V[\vec{R}/\vec{X}]$ by hypothesis.

---

[1] Can be shown by an easy induction on v.

Since $(\lambda x.v)u$ is neutral and it converts to reducible terms only, it is reducible. Hence $\lambda x.v$ is reducible. $\square$

Remark 3.1.14 *Working only with terms in* gentop *normal form allows us to rule out all the other reductions that are possible when considering all the terms of the calculus. This restriction is essential since otherwise we ought now to face, in Lemma 3.1.12, reductions like* $p_1\langle rep(U_1), p_2 w\rangle \longrightarrow p_1 w$, *that we cannot handle, for nothing in our induction hypothesis allows us to conclude that* $p_1 w$ *is reducible. (We already pointed out the difficulty in Section 2.3.2.) This reduction[2] is now ruled out as* $p_1\langle rep(U_1), p_2 w\rangle$ *is not a gentop normal form (its normal form being* $rep(U_1)$). *Similarly, in Lemma 3.1.13, the restriction to terms in* gentop *normal form allows us to rule out (in the case* $U \in Iso(\mathbf{T})$) *all the other reductions otherwise possible in the full calculus. As pointed out in the introduction (Section 2.3.2), we do not know how to handle the general reduction* $(\lambda x.(v'\,rep(U)))u \longrightarrow (v'u)$ *via* η_{top}: *if* u *is not* $rep(U)$, *then we have nothing in our induction hypothesis to tell us that* $(v'u)$ *is reducible. But here* u *must be in* gentop *normal form, that is to say,* $u = rep(U)$, *and the* η_{top} *reduction can be handled as above.*

Lemma 3.1.15 *(Universal abstraction) If for every type* V *and candidate* S *of type* V, $v[V/Y] \in RED_W[\overrightarrow{R}/\overrightarrow{X}, S/Y]$, *then* $\lambda Y.v \in RED_{\forall Y.W}[\overrightarrow{R}/\overrightarrow{X}]$.

Proof. We need to show that $(\lambda Y.v)[V] \in RED_W[\overrightarrow{R}/\overrightarrow{X}, S/Y]$ for every type V and candidate S of type V. We argue by induction on $\nu(v)$, using the fact that $(\lambda Y.v)[V]$ is neutral. Converting a redex of $(\lambda Y.v)[V]$ can yield:

- $(\lambda Y.v')[V]$ with v' one step from v; now, by induction hypothesis on $\nu(v)$, we know that $(\lambda Y.v')[V] \in RED_W[\overrightarrow{R}/\overrightarrow{X}, S/Y]$.

The result follows by CR3. $\square$

Lemma 3.1.16 $RED_{T[V/Y]}[\overrightarrow{R}/\overrightarrow{X}] = RED_T[\overrightarrow{R}/\overrightarrow{X}, RED_V[\overrightarrow{R}/\overrightarrow{X}]/Y]$

Proof. By induction on T. $\square$

Lemma 3.1.17 *(Universal application) If* $t \in RED_{\forall Y.W}[\overrightarrow{R}/\overrightarrow{X}]$, *then* $t[V] \in RED_{W[V/Y]}[\overrightarrow{R}/\overrightarrow{X}]$ *for every type* V.

Proof. By hypothesis, $t[V] \in RED_W[\overrightarrow{R}/\overrightarrow{X}, S/Y]$ for every candidate S. Taking $S = RED_V[\overrightarrow{R}/\overrightarrow{X}]$, the result follows by Lemma 3.1.16. $\square$

---

[2]And its symmetric $p_2\langle p_1 w, rep(U_2)\rangle \longrightarrow p_2 w$.

The theorem

As in [GLT90], we say here that a term t of type T is *reducible* if it is in $RED_T[\overrightarrow{SN}/\overrightarrow{X}]$, where $\overrightarrow{X}$ are the free type variables of T and SN_i is the set of strongly normalizable terms of type X_i. In the proof of the theorem, there is the need of a stronger induction hypothesis, from which the strong normalization follows by putting $u_i = x_i$ and $R_i = SN_i$.

Proposition 3.1.18 *Let* $t{:}T$ *be any term of* $\lambda^2\beta\eta\pi*$ *(in gentop normal form), whose free variables are among* $x_1 : U_1, \cdots, x_n : U_n$, *and all the free variables of* T, U_1, $\cdots U_n$ *are among* X_1, $\cdots X_m$. *If* $R_1, \ldots R_m$ *are reducibility candidates of types* $V_1, \cdots V_m$, *and* $u_1, \cdots, u_m$ *are terms of types* $U_1[\overrightarrow{V}/\overrightarrow{X}], \ldots U_m[\overrightarrow{V}/\overrightarrow{X}]$ *which are in* $RED_{U_1}[\overrightarrow{R}/\overrightarrow{X}], \ldots, RED_{U_n}[\overrightarrow{R}/\overrightarrow{X}]$, *then* $t[\overrightarrow{V}/\overrightarrow{X}][\overrightarrow{u}/\overrightarrow{x}] \in RED_T[\overrightarrow{R}/\overrightarrow{X}]$.

Proof. By induction on t. Notice that there are no variables of type U if $U \in Iso(\mathbf{T})$.

- $t = *$: t is in the only reducibility candidate $\{*\}$ of type $\mathbf{T}$.

- $t = x_i$: in this case the statement of the theorem becomes a tautology.

- $t = p_i u$: then $u : U_1 \times U_2$ and $U_i \notin Iso(\mathbf{T})$ as we consider only terms in *gentop* normal form. By induction hypothesis, $u[\overrightarrow{V}/\overrightarrow{X}][\overrightarrow{u}/\overrightarrow{x}] \in RED_{U_1 \times U_2}[\overrightarrow{R}/\overrightarrow{X}]$. Hence $(p_i u)[\overrightarrow{V}/\overrightarrow{X}][\overrightarrow{u}/\overrightarrow{x}] = p_i u[\overrightarrow{V}/\overrightarrow{X}][\overrightarrow{u}/\overrightarrow{x}] \in RED_{U_i}[\overrightarrow{R}/\overrightarrow{X}]$ by definition of reducibility for product types.

- $t = \langle u, v \rangle$: $u[\overrightarrow{V}/\overrightarrow{X}][\overrightarrow{u}/\overrightarrow{x}] \in RED_{U_1}[\overrightarrow{R}/\overrightarrow{X}]$ and $v[\overrightarrow{V}/\overrightarrow{X}][\overrightarrow{u}/\overrightarrow{x}] \in RED_{U_2}[\overrightarrow{R}/\overrightarrow{X}]$ by the induction hypothesis, so Lemma 3.1.12 gives $\langle u[\overrightarrow{V}/\overrightarrow{X}][\overrightarrow{u}/\overrightarrow{x}], v[\overrightarrow{V}/\overrightarrow{X}][\overrightarrow{u}/\overrightarrow{x}] \rangle \in RED_{U_1 \times U_2}[\overrightarrow{R}/\overrightarrow{X}]$. But, by definition, $\langle u, v \rangle[\overrightarrow{V}/\overrightarrow{X}][\overrightarrow{u}/\overrightarrow{x}]$ is $\langle u[\overrightarrow{V}/\overrightarrow{X}][\overrightarrow{u}/\overrightarrow{x}], v[\overrightarrow{V}/\overrightarrow{X}][\overrightarrow{u}/\overrightarrow{x}] \rangle$, and hence $\langle u, v \rangle[\overrightarrow{V}/\overrightarrow{X}][\overrightarrow{u}/\overrightarrow{x}] \in RED_{U_1 \times U_2}[\overrightarrow{R}/\overrightarrow{X}]$.

- $t = \lambda z.v$: by induction hypothesis, we know that $v[\overrightarrow{V}/\overrightarrow{X}][\overrightarrow{u}/\overrightarrow{x}][u/z] \in RED_V[\overrightarrow{R}/\overrightarrow{X}]$ for all $u \in RED_U[\overrightarrow{R}/\overrightarrow{X}]$. Then Lemma 3.1.13 gives $\lambda z.v[\overrightarrow{V}/\overrightarrow{X}][\overrightarrow{u}/\overrightarrow{x}] \in RED_{U \to V}[\overrightarrow{R}/\overrightarrow{X}]$. But $(\lambda z.v)[\overrightarrow{V}/\overrightarrow{X}][\overrightarrow{u}/\overrightarrow{x}]$ is $\lambda z.v[\overrightarrow{V}/\overrightarrow{X}][\overrightarrow{u}/\overrightarrow{x}]$ by definition, and the result follows.

- $t = vu$: then $v[\overrightarrow{V}/\overrightarrow{X}][\overrightarrow{u}/\overrightarrow{x}] \in RED_{U \to V}[\overrightarrow{R}/\overrightarrow{X}]$ and by induction hypothesis $u[\overrightarrow{V}/\overrightarrow{X}][\overrightarrow{u}/\overrightarrow{x}] \in RED_U[\overrightarrow{R}/\overrightarrow{X}]$. Hence we know that $(v[\overrightarrow{V}/\overrightarrow{X}][\overrightarrow{u}/\overrightarrow{x}]\ u[\overrightarrow{V}/\overrightarrow{X}][\overrightarrow{u}/\overrightarrow{x}]) \in RED_V[\overrightarrow{R}/\overrightarrow{X}]$, as it is $(vu)[\overrightarrow{V}/\overrightarrow{X}][\overrightarrow{u}/\overrightarrow{x}]$ by definition.

- $t = \lambda Y.v$: then we know by induction hypothesis that for every type V and reducibility candidate S we have $v[V/Y][\overrightarrow{V}/\overrightarrow{X}][\overrightarrow{u}/\overrightarrow{x}] \in RED_W[\overrightarrow{R}/\overrightarrow{X}, S/Y]$. Then, applying Lemma 3.1.15, we get that $(\lambda Y.v)[\overrightarrow{V}/\overrightarrow{X}][\overrightarrow{u}/\overrightarrow{x}] \in RED_{\forall Y.W}[\overrightarrow{R}/\overrightarrow{X}]$.

- $t = t[V]$: then we know by induction hypothesis that $t[\overrightarrow{V}/\overrightarrow{X}][\overrightarrow{u}/\overrightarrow{x}] \in RED_{\forall Y.W}[\overrightarrow{R}/\overrightarrow{X}]$ and Lemma 3.1.17 allows to conclude that for every type V $t[V][\overrightarrow{V}/\overrightarrow{X}][\overrightarrow{u}/\overrightarrow{x}] \in RED_{W[V/Y]}[\overrightarrow{R}/\overrightarrow{X}]$.

□

Theorem 3.1.19 $\overset{\beta\eta^2\pi*}{\longrightarrow}$ *is strongly normalizing over the set of* gentop *normal forms.*

Proof. Let t be any term in *gentop* normal form. All its free variables are in any reducibility candidate by CR3, so that $t = t[\overrightarrow{SN}/\overrightarrow{X}][\overrightarrow{x}/\overrightarrow{x}]$ is reducible by the previous lemma. By CR1 it is strongly normalizing. That is, $\overset{\beta\eta^2\pi*}{\longrightarrow}$ is strongly normalizing over *gentop* normal forms. □

3.2   Normalization without η_{top} and SP_{top}

The proof of strong normalization is essentially the same as the one given above for the full system without β^2 over the subset of terms in *gentop* normal form.

The main difference, besides the fact that we add β^2 and *gentop* and exclude η_{top} and SP_{top}, is that now we work on the full set of terms, so that there are plenty of terms $t{:}U$, besides $rep(U)$, when $U \in Iso(\mathbf{T})$. We keep essentially the same notion of neutral term (3.1.1), but it is to be noted that only $rep(U)$ is neutral, not every term of type $U \in Iso(\mathbf{T})$.

Definition 3.2.1 (neutral terms) *A term $t : U$ is neutral iff at least one of the following conditions is satisfied:*

- *if $U \notin Iso(\mathbf{T})$ and t is not an abstraction, a type abstraction or a pair,*

- *if $U \in Iso(\mathbf{T})$ and t is rep(U).*

Since we drop η_{top} and SP_{top}, there is no need to give a special status to the types $U \in Iso(\mathbf{T})$ (besides the fact that $rep(U)$ is neutral), and we resort to the usual definition of product and function space of reducibility candidates, which allows us to deal with all the terms of type $U \in Iso(\mathbf{T})$.

Definition 3.2.2 (product and arrow of reducibility candidates) *If R and S are reducibility candidates of types U and V, we define:*

- $t \in R \to S \iff$ for all $u \in R$, $tu \in S$

- $t \in R \times S \iff \mathrm{p_1}t \in U$ and $\mathrm{p_2}t \in V$

With this new definition, the proofs of the previous appendix go through almost unchanged, with the only care to keep in mind that now $rep(U)$ is no longer the only term of type $U \in Iso(\mathbf{T})$, and that types in $Iso(\mathbf{T})$ have no longer a special status. This means that wherever there is a distinction between types that are in $Iso(\mathbf{T})$ and types that are not, one follows the proof given for types that are not in $Iso(\mathbf{T})$. The new cases arising from *gentop* reductions are easily dealt with, as $rep(U)$ is still in any reducibility candidate by CR3.

For completeness, we detail here all the changes that are needed.

- Remark 3.1.4 now extends to *all* variables and also the variables of type $U \in Iso(\mathbf{T})$. It is just the matter of noticing that a variable $x{:}U \in Iso(\mathbf{T})$ is neutral and reduces only to $rep(U)$, that is, in any reducibility candidate by CR3, and the result follows by CR3.

- In Theorem 3.1.8, we can no longer factor out the types in $Iso(\mathbf{T})$, that must be treated exactly as the other types:

  Product Types (CR3)

  * t can be $rep(U_1 \times U_2)$. In that case the only possible reduction for $\mathrm{p}_i t$ (that is not in *gentop* normal form) is to $rep(U_i)$, that belongs to all reducibility candidate (Remark 3.1.4), hence in $RED_{U_i}[\overrightarrow{R}/\overrightarrow{X}]$ that is a reducibility candidate by induction hypothesis on U_i. So $\mathrm{p}_i t \in RED_{U_i}[\overrightarrow{R}/\overrightarrow{X}]$ by CR3 on U_i and we get $t \in RED_{U_1 \times U_2}[\overrightarrow{R}/\overrightarrow{X}]$ by definition.
  * t can be a neutral term different from $rep(U_1 \times U_2)$. Then the only possible reduction for $\mathrm{p}_i t$ (that is not in *gentop* normal form) is to $rep(U_i)$, and we conclude as above.

  Arrow Types (CR3)

* t (or t') can be $rep(U_1 \rightarrow U_2)$. Then (tu) (or $(t'u)$) can only reduce to $rep(U_2)$ that is in any reducibility candidate (Remark 3.1.4), hence in $RED_{U_2}[\overrightarrow{R}/\overrightarrow{X}]$ that is a reducibility candidate by induction hypothesis on U_2. So (tu) (or $(t'u)$) $\in RED_{U_2}[\overrightarrow{R}/\overrightarrow{X}]$ for all $u \in RED_{U_1}[\overrightarrow{R}/\overrightarrow{X}]$, and we get $t \in RED_{U_1 \rightarrow U_2}[\overrightarrow{R}/\overrightarrow{X}]$ by definition.

* t can be a neutral term different from $rep(U_1 \rightarrow U_2)$. Then the only possible reduction for (tu) (or $(t'u)$) is to $rep(U_2)$, and we conclude as above.

- In Theorem 3.1.11, we can no longer factor out the types in $Iso(\mathbf{T})$ that must be treated exactly as the other types.

Universal Types (CR3)

* t (or t') can be $rep(\forall Y.W)$. Then $t[V]$ can only reduce to $rep(W)$, that is in any reducibility candidate (Remark 3.1.4), hence in $RED_W[\overrightarrow{R}/\overrightarrow{X}]$ that belongs to all reducibility candidate by induction hypothesis on W. Again we get t(or t') $\in RED_{\forall Y.W}[\overrightarrow{R}/\overrightarrow{X}]$ by definition.

* t (or t') can be a neutral term different from $rep(\forall Y.W)$. Then $t[V]$ can only reduce to $rep(W)$, and we conclude as above.

- In Lemmas 3.1.12 and 3.1.13 we can no longer factor out the case of types $U \in Iso(\mathbf{T})$, which must be treated uniformly as the other types. Since the rules SP_{top} and η_{top} are not present, only the first four cases considered in Lemma 3.1.12 can occur, and the proof goes through unchanged for them, while for Lemma 3.1.13 we follow the proof given for $V \notin Iso(\mathbf{T})$.

  There is now the further possibility of a *gentop* reduction, that in both cases is dealt with in the usual way by remembering that *any* reducibility candidate of type $U \in Iso(\mathbf{T})$ contains $rep(U)$.

- In Lemma 3.1.15 we have now two additional cases:

  - $(\lambda Y.v)[V]$ reduces to the term $rep(\mathrm{W}[V/Y])$, that must belong to $RED_{W[V/Y]}[\overrightarrow{R}/\overrightarrow{X}]$ since this latter is a reducibility candidate.
  - $(\lambda Y.v)[V]$ reduces to $v[V/Y]$. But we know by hypothesis that $v[V/Y] \in RED_{W[V/Y]}[\overrightarrow{R}/\overrightarrow{X}, S/Y]$

- In the proof of the Proposition 3.1.18, it suffices to apply to the types $V \in Iso(\mathbf{T})$ the same arguments used for types $U \notin Iso(\mathbf{T})$, as now there is no longer any difference in the definition of the function space and product of reducibility candidates.

Using again the fact that $t = t[\overrightarrow{SN}/\overrightarrow{X}][\overrightarrow{x}/\overrightarrow{x}]$, we similarly get our final result.

Theorem 3.2.3 $\overset{\beta^2\eta^2\pi*}{\longrightarrow}$ *without* η_{top} *and* SP_{top} *is strongly normalizing.*

Chapter 4

First-Order Isomorphic Types

In this chapter we characterize the isomorphisms which hold in all models of $\lambda^1\beta\eta\pi*$, the simply typed lambda calculus with surjective pairing (and "terminal object"). Moreover, we show that it is decidable whether two types (built from type variables) are isomorphic in all models of this calculus. It is well known that these models are exactly the Cartesian Closed Categories (ccc).

The main theorem of this chapter shows that two types A and B can be constructively proved to be isomorphic, by two programs which act one as the inverse of the other, iff $Th^1_{\times\mathbf{T}} \vdash A = B$.

Let us recall the definition of $Th^1_{\times\mathbf{T}}$.

Definition 4.0.4 $Th^1_{\times\mathbf{T}}$ is a theory of equality plus the following axiom schemas, where $\mathbf{T}$ is a constant symbol:

1. $A \times B = B \times A$

2. $A \times (B \times C) = (A \times B) \times C$

3. $(A \times B) \to C = A \to (B \to C)$

4. $A \to (B \times C) = (A \to B) \times (A \to C)$

5. $A \times \mathbf{T} = A$

6. $A \to \mathbf{T} = \mathbf{T}$

7. $\mathbf{T} \to A = A$

In order to discuss the soundness of $Th^1_{\times\mathbf{T}}$ and explain where it comes from, we hint here of its categorical meaning. Note, though, that no notion nor result from category theory is used in this chapter.

Since models of the typed lambda calculus with surjective pairing are exactly the Cartesian Closed Categories (ccc), our results translate directly into theorems on when two generic objects are isomorphic in all ccc's. In other words, $Th^1_{\times\mathbf{T}}$ characterizes the objects which are isomorphic just by the cartesian closed structure of the category in which they are interpreted, no matter which particular ccc is chosen.

Observe first that $Th^1_{\times\mathbf{T}}$ is realized in every Cartesian Closed Category, when "=" is interpreted as isomorphism. The first three axioms describe properties of the cartesian product (associativity, commutativity, identity for $\times$), and the second three axioms can be seen as the properties of the three adjunctions of a ccc that relate product, exponent and the terminal object. The last equation ($\mathbf{T} \to A = A$) tells us that the arrows from the terminal object to A in a ccc are the points of A. Thus, the theory $Th^1_{\times\mathbf{T}}$ is sound. A consequence of our main result is the completeness of $Th^1_{\times\mathbf{T}}$

with respect to ccc's. That is, no other isomorphism is valid in all ccc's. (This is not obvious because there are categorical models of $Th^1_{\times T}$ which are not ccc's: take a Cartesian Category with a bifunctor " $\to$ " such that $A \to B = B$, say.)

A further consequence of the work below in λ-calculus will be an insight into the composition of derivations in proof theory. The typed lambda calculus with surjective pairing is the language for proofs of $IPC(\mathbf{True}, \wedge, \to)$, the intuitionistic positive propositional calculus. In the proof theoretic framework we then characterize *strongly equivalent formulae*, where two formulae A and B are considered equivalent if, given a proof f of the sequent $A \vdash B$, and a proof g of the sequent $B \vdash A$, $g\, f$ yields, after cut-elimination, the identity proof of the sequent $A \vdash A$ and vice-versa. The details of both the categorical and proof-theoretic applications are discussed in [DCL91].

The chapter is organized as follows. Section 4.1 sets out the basic definitions leading up to the notion of a type normal form. Section 4.2 presents some rather technical lemmas which will be used in Section 4.3 in order to characterize the set of provable isomorphisms and to show the decidability of the theory $Th^1_{\times T}$.

4.1 Rewriting types

This section, as well as Sections 4.2 and 4.3, is dedicated to the proof of the *completeness* of $Th^1_{\times T}$ for $\lambda^1 \beta\eta\pi*$. The first steps are done by reducing types to a "type normal form" that is very similar to the one used by Hindley in [Hin82]. The axioms of $Th^1_{\times T}$ suggest the following rewrite system $\mathcal{R}^1$ for types (essentially $Th^1_{\times T}$ without commutativity):

Definition 4.1.1 (Type rewriting $\mathcal{R}^1$) *Let "$\mathcal{R}^1$" be the transitive and substitutive* **type-reduction** *relation generated by:*

1. $A \times (B \times C) > (A \times B) \times C$

2. $(A \times B) \to C > A \to (B \to C)$

3. $A \to (B \times C) > (A \to B) \times (A \to C)$

4. $A \times \mathbf{T} > A$

5. $\mathbf{T} \times A > A$

6. $A \to \mathbf{T} > \mathbf{T}$

7. $\mathbf{T} \to A > A$

The system $\mathcal{R}^1$ yields an obvious notion of **normal form for types** (type-n.f.), i.e., when no type reduction is applicable. Note that 5, 6 and 7 "eliminate the **T**'s", while 2 and 3 "bring the $\times$ outside". It is then easy to observe that each type-n.f. is **T** or has the structure $B_1 \times \cdots \times B_n$ where each B_i does not contain **T** or "$\times$". We write *nf(B)* for the normal form of B (there is exactly one, see 4.1.2), and say that a normal form is nontrivial if it is not **T**.

Proposition 4.1.2 *Each type has a unique type normal form in* $\mathcal{R}^1$.

Proof. Notice that in any $\mathcal{R}^1$-reduction, starting with a given type B:

(i) Rules 2 and 3 can be applied only finitely many times, as they strictly decrease the number of $\times$'s in the scope of an arrow of B and this number is finite and is not increased by any other rule.

(ii) Between an application of rule 2 or 3 (yielding type B') and the next one, the remaining rules can be applied only finitely many times (5, 6, 8 and 7 simply throw away some subformula reducing by one the number of products or arrows, which is finite; rule 1 is just associativity to the left).

So, after a finite reduction path we get a type B'' with no redex for rules 2 and 3, and then, again, the remaining rules can be applied only finitely many times (at most the length of B'' plus the times required for associating B'' to the left). The resulting type nf(B) has then no products in the scope of any arrow (otherwise 2 and 3 could be applied), and is either **T** or a type with no occurrence of **T** (otherwise 5, 6, 8 and 7 could be applied). Thus nf(B) is a product of types, each of which has no occurrence of $\times$.

It is easy to observe that $\mathcal{R}^1$ is Church-Rosser too and, thus, that nf(B) is unique. (Note also that we have actually proved that $\mathcal{R}^1$ strongly normalizes). $\square$

From the implication proved above of the main theorem, since $\mathcal{R}^1 \vdash B > A$ implies $Th^1_{\times \mathbf{T}} \vdash B = A$, it is clear that any reduction $\mathcal{R}^1 \vdash B > A$ is witnessed (or, proved, in the "types-as-propositions" analogy) by an invertible term of type $B \to A$. Moreover, one clearly has:

Corollary 4.1.3 $Th^1_{\times \mathbf{T}} \vdash B = nf(B)$ *and, thus,* $Th^1_{\times \mathbf{T}} \vdash B = A \iff Th^1_{\times \mathbf{T}} \vdash nf(B) = nf(A)$.

In conclusion, when $Th^1_{\times \mathbf{T}} \vdash B = A$, either we have $nf(B) \equiv \mathbf{T} \equiv nf(A)$, or $Th^1_{\times \mathbf{T}} \vdash nf(B) \equiv B_1 \times \cdots \times B_n = A_1 \times \cdots \times A_m \equiv nf(A)$. A crucial lemma below will prove that, in this case, one also has $n = m$.

The assertion in the corollary can be reformulated for invertible terms in a very convenient way:

Proposition 4.1.4 (commuting diagram) *Given types A and B, assume that the invertible terms $F{:}A \to nf(A)$ and $G{:}B \to nf(B)$ prove the reductions to type-normal-form. Then a term $M{:}A \to B$ is invertible iff there exists an invertible term $M'{:}nf(A) \to nf(B)$, such that $M = G^{-1} \circ M' \circ F$*

Proof. ($\Leftarrow$) Set $M^{-1} \equiv (G^{-1}M' \circ F\circ)^{-1} \equiv F^{-1} \circ M'^{-1} \circ G$, then M is invertible.

($\Rightarrow$) Just set $M' \equiv G \circ M \circ F^{-1}$. Then $M'^{-1} \equiv F \circ M^{-1} \circ G^{-1}$ and M' is invertible. $\square$

The diagram in Figure 4.1 represents the situation in the corollary.

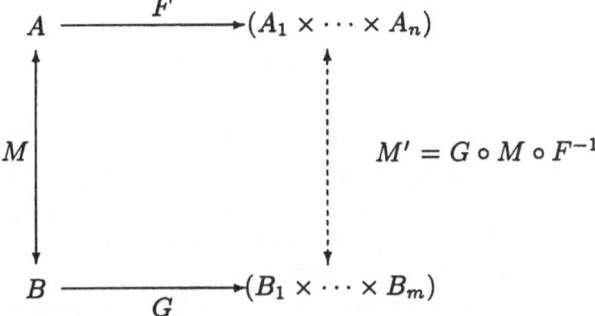

Figure 4.1: Reduction to a subclass of isomorphic types.

Thus we have reduced isomorphisms between arbitrary types to the same problems with respect to type normal forms. We examine next how this may affect the structure of the terms which prove the isomorphisms.

4.2 From $\lambda^1\beta\eta\pi*$ to the classical $\lambda^1\beta\eta$

This is a technical section, where we prove some crucial lemmas. Our aim is to reduce invertibility in $\lambda^1\beta\eta\pi*$ to invertibility in $\lambda^1\beta\eta$.

Recall first that, when $Th^1_{\times\mathbf{T}} \vdash B = A$, one has $nf(B) \equiv \mathbf{T} \equiv nf(A)$, or $Th^1_{\times\mathbf{T}} \vdash nf(B) \equiv B_1 \times \cdots \times B_n = A_1 \times \cdots \times A_m \equiv nf(A)$. Notice now that, in the latter case, there cannot be any occurrence of $\mathbf{T}$ in either type. Indeed, a nontrivial type-n.f. cannot be provably equated to $\mathbf{T}$, as can be easily seen by taking a nontrivial model. Thus we restrict our attention to equations like $B_1 \times \cdots \times B_n = A_1 \times \cdots \times A_m$ with no occurrence of $\mathbf{T}$ and, hence, to invertible terms with no occurrence of the type constant $\mathbf{T}$ in their types. We can show that these terms do not contain any occurrence of $*$ either, for any type A, via the following lemmas.

Lemma 4.2.1 (form of the terms of a product type) *Given a term M of $\lambda^1\beta\eta\pi*$ in normal form such that $M : A \times B$, then either $M \equiv \langle M_1, M_2 \rangle$, for some M_1, M_2, or there is a free variable $x : C$ in M such that $A \times B$ is a type subexpression of C.*

Proof. By induction on the length of the structure of M.

Basis of induction: if M is of length 1, then it can be only a free variable of type $A \times B$.

Inductive step: $M \equiv \lambda\overrightarrow{x}.r\overrightarrow{P}$, as it is in normal form. Observe first that this case reduces to $M \equiv r\overrightarrow{P}$, as its type is $\alpha \times \beta$, and we proceed by case analysis on r as follows:

  r is a variable: then r is free and has type
    $type(P_1) \rightarrow (\cdots \rightarrow (type(P_n) \rightarrow A \times B)\cdots)$.

  r is $\langle M_1, M_2 \rangle$: then $M \equiv \langle M_1, M_2 \rangle$, in order to type-check.

  r is p_1 or p_2: then $M \equiv (\cdots (p_i M_1) M_2 \cdots M_k)$ with $M_1:S \times U$ in normal form with S or $U \equiv type(M_2) \rightarrow (\cdots \rightarrow (type(M_k) \rightarrow A \times B)\cdots)$. By induction hypothesis either M_1 is $\langle N_1, N_2 \rangle$ or M_1 has a free variable $x : C$ with $S \times U$ (hence $A \times B$ too) a type subexpression of C. The first case is not possible, as $p_i\langle N_1, N_2 \rangle$ is a redex, so M has a free variable $x : C$ with $A \times B$ a type subexpression of C.

  r is $*$: this is not possible as $*$ has type **T**, which would prevent M from having type $A \times B$.

□

Lemma 4.2.2 (no $*$ in a term in n.f. if no T in its type)
*Assume that in a term M of $\lambda^1\beta\eta\pi*$ in normal form there is an occurrence of $*$. Then there is some occurrence of the type constant **T** in the type of M or in the type of some free variable of M.*

Proof. By induction on the structure of M.

Basis for induction: $*$ has type **T**.

Inductive step: $M \equiv \lambda\overrightarrow{x}.r\overrightarrow{P}$, as M is in normal form, and we proceed by case analysis on r as follows:

  r is a variable: then r has type $type(P_1) \rightarrow (\cdots \rightarrow (type(P_n) \rightarrow C)\cdots)$; by hypothesis, the P_i's are in normal form and in some P_j

there are occurrences of a constant $*$, so by induction hypothesis
there are **T**'s in $type(P_j)$, and hence in the type of r. By this,
either r is a free variable or (since r occurs among the $\overrightarrow{x}$) there
are **T**'s in the type of M.

r **is** $\langle P, Q \rangle$: then $M \equiv \lambda \overrightarrow{x}.\langle P, Q \rangle$ where P and Q are in normal form.
The type of M is $D_1 \to \cdots \to D_n \to (A \times B)$, with $P : A$ and
$Q : B$, and $*$ occurs in P or Q. By inductive hypothesis, either
T occurs in $A \times B$ (hence in the type of M, too) or in the type
of some free variable y of P or Q. In either case, as above, some
T's occur in the type of M or in the type of y, which is free in
M.

r **is** p_1 **or** p_2: then $M \equiv \lambda \overrightarrow{x}.((\mathrm{p}_i M_1) M_2 \cdots M_k)$ where:

- M_j is in normal form, for each j.
- $M_1 : S \times U$ with either S or $U \equiv type(M_2) \to (\cdots \to (type(M_k) \to C) \cdots)$.
- $*$ occurs in M_j for some j; consider than

   case $j = 1$: then **T** occurs in $S \times U$, by induction hypothesis. By Lemma 4.2.1, as M cannot be a redex, M_1 is
   not a pair and has a free variable $y : C$ with $S \times U$ a
   type subexpression of C. Notice that y is also free in
   $((\mathrm{p}_i M_1) M_2 \cdots M_k)$. Thus as in the earlier cases either y
   is free in M or some **T**'s occur in the type of M (because
   y is one of the variables in $\overrightarrow{x}$);

   case $j > 1$: then by induction hypothesis either

   (a) there is a **T** occurring in the type of M_j, and hence
   in $S \times U$, or

   (b) there is a free variable y of M_j with type **T** occurring
   in its type.

   In case (a), we can conclude the proof as in the case for
   $j = 1$ above. In case (b), if y is free in M_j then it is
   also free in $((\mathrm{p}_i M_1) M_2 \cdots M_k)$. We can thus conclude
   the proof again as for $i = 1$.

r **is** $*$: then $M \equiv \lambda \overrightarrow{x}.*$ and the type of M is $D_1 \to \cdots D_n \to \mathbf{T}$, for
some $D_1, \ldots, D_n$.

$\square$

Proposition 4.2.3 (bijections between type-n.f.'s are in $\lambda^1\beta\eta\pi$**)**
*Assume that B and A are nontrivial type-n.f.'s. If the closed terms M and
N prove $B \cong_d A$ in $\lambda^1\beta\eta\pi*$, then their normal forms contain no occurrences
of the constants $*$. (Thus, M and N are actually in $\lambda^1\beta\eta\pi$.)*

Proof. By the previous lemma, since the terms are closed and no **T** occurs in their type. $\square$

So we have factored out the first class of constants $*$, and we have restricted ourselves to $\lambda^1\beta\eta\pi$. In the next step we eliminate pairing as well, in a sense.

There is a problem though. Our aim is to reduce the investigation of invertible terms in $\lambda^1\beta\eta\pi*$ to that of terms in $\lambda^1\beta\eta\pi$. This is done on the grounds of Proposition 4.1.4 by examining each component of the product, where the isomorphism will be given by terms of $\lambda^1\beta\eta$. However, in the notation of Proposition 4.1.4, consider the term $M' : nf(A) \rightarrow nf(B)$. M' is invertible in (the equational theory of) $\lambda^1\beta\eta\pi*$ and thus also the subterms yielding the isomorphism of the components (see 4.2.7 and 4.2.8 below) are *a priori* invertible in $\lambda^1\beta\eta\pi*$, while we need to know that they are actually invertible in $\lambda^1\beta\eta$. We get rid of the problem by the following remark.

Remark 4.2.4 *(The equational theory of)* $\lambda^1\beta\eta\pi*$ *is a conservative extension of (the equational theory of)* $\lambda^1\beta\eta$. *Similarly for* $\lambda^1\beta\eta\pi$ *with respect to* $\lambda^1\beta\eta$.

Indeed, this is a simple application of Corollary 2.6.2 from Chapter 2.

Notation 4.2.5 Recall that by $\overrightarrow{x}$, $\overrightarrow{y}$, $\overrightarrow{M}$, ... we denote vectors of variables, terms, etc.

Lemma 4.2.6 (terms of $\lambda^1\beta\eta\pi$ with arrow-only type are in $\lambda^1\beta\eta$)
Let M be a term of $\lambda^1\beta\eta\pi$ in normal form such that $M : A$, where A is a type with no occurrence of $\times$ in it. If no free variable of M has a type with occurrences of $\times$, then M is actually a term in $\lambda^1\beta\eta$.

Proof. By induction on the structure of M.

Basis for induction: if M is of length 1, then it can be only a variable of type A, as any constant has a type with occurrences of $\times$.

Inductive step: $M \equiv \lambda\overrightarrow{x}.r\overrightarrow{P}$, as M is in normal form, and we proceed by case analysis on r as follows:

   r **is a variable:** then r has type $type(P_1) \rightarrow (\cdots \rightarrow (type(P_n) \rightarrow C)\cdots)$ and no matter if r is free or bound, by hypothesis on the type of M and its free variables, the P_i's (which are in normal form) have a type with no $\times$'s and free variables whose type have no $\times$'s, so by induction hypothesis they contain no constants nor pairs, and hence M contains no constants or pairs either.

r is $\langle P, Q\rangle$: this is impossible, otherwise $M \equiv \lambda\vec{x}.\langle P,Q\rangle$ and the type of M would be $S_1 \to \cdots \to S_n \to (A \times B)$, which contains $\times$.

r is p_1 or p_2: this cannot be either, since:

- $M \equiv \lambda\vec{x}.p_i$ must have a type containing $\times$,
- $M \equiv \lambda\vec{x}.((p_i M_1)M_2\cdots M_k)$ implies, by lemma 4.2.1, that either M_1 is $\langle N_1, N_2\rangle$ or M_1 has a free variable $x : C$ with $S\times U$ a type subexpression of C. The first case is not possible, as $p_i\langle N_1, N_2\rangle$ is a redex while M is in normal form. Thus M_1 has a free variable $x : C$ with $S \times U$ a type subexpression of C, and hence either $x \in FV(M)$ or $S \times U$ is a type subexpression of the type of M, since the type of M includes the types of bound variables. Impossible.

□

Proposition 4.2.7 (isolate the relevant $\langle M_1, \cdots, M_n\rangle$)
Let $B \equiv B_1 \times \cdots \times B_m$ and $A \equiv A_1 \times \cdots \times A_n$ be type-n.f.'s where neither the B_i's nor the A_j's contain any occurrences of $\mathbf{T}$ or $\times$. Then $B \cong_p A$ iff there exist $M_1, \cdots, M_n$ and $N_1, \cdots, N_m$ such that

$$x_1 : B_1, \cdots, x_m : B_m \vdash M_1, \cdots, M_n \quad M_i[\vec{N}/\vec{x}] =_{\beta\eta} y_i, \text{ for } 1 \le i \le n$$

$$y_1 : A_1, \cdots, y_n : A_n \vdash N_1, \cdots, N_m \quad N_j[\vec{M}/\vec{y}] =_{\beta\eta} x_j, \text{ for } 1 \le i \le m$$

(where substitution of vectors of equal length is meant componentwise).

Proof.
$(\Rightarrow)$

Let $M^\circ: B \to A$ and $N^\circ : A \to B$ be closed terms (in normal form) of $\lambda^1\beta\eta\pi*$ such that $M^\circ \circ N^\circ = I_A$ and $N^\circ \circ M^\circ = I_B$. Then by standard currying, consider the term $\lambda x_1, \cdots, x_m.M^\circ < x_1, \cdots, x_m >: (B_1 \to \cdots \to (B_m \to (A_1 \times \cdots \times A_n), \cdots,),$ and observe that the normal form M' of $M^\circ < x_1, \cdots, x_m > : A_1 \times, \cdots, \times A_n$, by Lemma 4.2.1, must be of the form $< M_1, \cdots, M_n >$, with $FV(M') = \{x_1 : B_1,, \cdots, x_m : B_m\}$ (by assumption, the B_i's contain no occurrences of $\times$). The same applies for N.

As for the other properties, let

$$M'' \equiv \lambda z.(\lambda x_1, \cdots, x_m.M^\circ < x_1, \cdots, x_m >)(p_1 z), \cdots, (p_m z)$$

and

$$N'' \equiv \lambda z.(\lambda y_1, \cdots, y_n.N^\circ < y_1, \cdots, y_n >)(p_1 z), \cdots, (p_n z),$$

where the x_i's, y_j's, and z are chosen to be distinct.
Then

$$M'' =_\beta \lambda z.M^\circ < p_1 z, \cdots, p_m z >=_\eta \lambda z.M^\circ z =_\eta M^\circ,$$

and similarly

$$N'' =_\beta \lambda z.N^\circ z =_\eta N^\circ.$$

Compute then

$$M^\circ \circ N^\circ =_{\beta\eta}$$

$$=_{\beta\eta} \quad M'' \circ N'' \equiv \lambda x.(M''(N''x)) \; x \text{ not occurring in } M'' \text{ or } N''.$$

$$=_{\beta\eta} \quad \lambda x.(\lambda z.(\lambda x_1, \cdots, x_m.M')(p_1 z), \cdots, (p_m z))(N''x)$$

$$=_{\beta\eta\pi} \quad \lambda x. < M_1[\overrightarrow{p_j(N''x)/\overrightarrow{x_j}}], \cdots, M_n[\overrightarrow{p_j(N''x)/\overrightarrow{x_j}}] >$$

(substitution done simultaneously for all $1 \leq j \leq m$)

$$=_{\beta\eta\pi} \quad \lambda x. < M_1[\overrightarrow{N}[\overrightarrow{p_i t}/\overrightarrow{y}]/\overrightarrow{x}], \cdots, M_n[\overrightarrow{N}[\overrightarrow{p_i t}/\overrightarrow{y}]/\overrightarrow{x}] >$$

since $N''x \quad =_{\beta\eta} \quad \lambda y_1, \cdots, y_n.N'(p_1 x) \cdots (p_n x)$

$$=_{\beta\eta} \quad < N_1[\overrightarrow{p_i t}/\overrightarrow{y}], \cdots, N_m[\overrightarrow{p_i t}/\overrightarrow{y}] >$$

(substitution done simultaneously for all $1 \leq i \leq n$)

$$=_{\beta\eta} \quad \lambda x. < M_1[\overrightarrow{N}/\overrightarrow{x}][\overrightarrow{p_i t}/\overrightarrow{y_i}], \cdots, M_n[\overrightarrow{N}/\overrightarrow{x}][\overrightarrow{p_i t}/\overrightarrow{y_i}] >$$

by substitution properties, as no y_i is free in M'

$$=_{\beta\eta\pi} \quad \lambda x. < p_1 x, \cdots, p_n x >$$

since $M^\circ \circ N^\circ =_{\beta\eta} \lambda x.x$ and $x =_{\beta\eta} < p_1 x, \cdots, p_n x >$.

Observe now that the equality just proved implies, componentwise, that
$M_k[\overrightarrow{N}/\overrightarrow{x}][\overrightarrow{p_i t}/\overrightarrow{y_i}] =_{\beta\eta\pi*} p_k x$ where $p_k x$ is a normal form because its type
B_k does not contain $\mathbf{T}$, so not even *gentop* is applicable to it. For the
purpose of the final argument of the proof, we refer now to $\overset{\beta\eta\pi*}{\rightarrow} {}^*$ as a
"several steps reduction" in $\lambda^1\beta\eta\pi*$. In view of the Church-Rosser Property
for this calculus, the last equality is equivalent to

$$M_k[\overrightarrow{N}/\overrightarrow{x}][\overrightarrow{p_i w}/\overrightarrow{y_i}] \overset{\beta\eta\pi*}{\rightarrow} {}^* p_k w,$$

where w is a fresh variable (to avoid confusion between $\overrightarrow{x}$ and x; in other
words, w is not free in M_k nor in any N_i and cannot be free in any reduct
of $M_k[\overrightarrow{N}/\overrightarrow{x}]$ either).

Notice now that by hypothesis the terms $\overrightarrow{M}$ and $\overrightarrow{N}$ are in normal form
and have no $\mathbf{T}$ or $\times$ involved in their types or in the types of their free

variables (the $\vec{B_i}$ and $\vec{A_i}$), so by Lemma 4.2.6 they are actually terms of $\lambda^1\beta\eta$. This allows us to conclude that the substitution $[\overrightarrow{p_i w}/\vec{y_i}]$ creates no new redexes: the $\overrightarrow{p_i w}$ could only create new redexes for surjective pairing reductions, i.e., when they appear in $< p_1 w, \cdots, p_n w >$. But $\vec{M}$ and $\vec{N}$ do not contain any pair, so surjective pairing reductions cannot apply.

This fact has an important consequence: the reductions are actually performed inside $M_k[\vec{N}/\vec{x}]$, so if we have $M_k[\vec{N}/\vec{x}][\overrightarrow{p_i w}/\vec{y_i}] \overset{\beta\eta\pi*}{\rightarrow} {}^*Q$, then $M_k[\vec{N}/\vec{x}] \overset{\beta\eta\pi*}{\rightarrow} {}^*Q'$ with $Q \equiv Q'[\overrightarrow{p_i w}/\vec{y_i}]$.

This implies, in the case of $M_k[\vec{N}/\vec{x}][\overrightarrow{p_i w}/\vec{y_i}] \overset{\beta\eta\pi*}{\rightarrow} {}^*p_k w$, the reduction $M_k[\vec{N}/\vec{x}] \overset{\beta\eta\pi*}{\rightarrow} {}^*Q'$ with $p_k w \equiv Q'[\overrightarrow{p_i w}/\vec{y_i}]$, that is $M_k[\vec{N}/\vec{x}] \overset{\beta\eta\pi*}{\rightarrow} {}^*p_k w$. In conclusion, $M_k[\vec{N}/\vec{x}] = y_k$, as required.

Symmetrically, one obtains $N_j[\vec{M}/\vec{y}] =_{\beta\eta} x_j$ from $N° \circ M° = \lambda x.x$.

($\Leftarrow$) Just step through the above proof in reverse order, defining the required closed terms by

$$M \equiv \lambda z.(\lambda x_1, \cdots, x_m. < M_1, \cdots, M_n >)(p_1 z), \cdots, (p_m z),$$

$$N \equiv \lambda z.(\lambda y_1, \cdots, y_n. < N_1, \cdots, N_m >)(p_1 z), \cdots, (p_n z).$$

□

In conclusion, we have isolated some interesting terms from which every constant has been factored out. Next we prove that provably equal types in normal form have equal length.

Lemma 4.2.8 (isomorphic type-n.f.'s have equal length)
Assume that $A_1 \times \cdots \times A_n$ *and* $B_1 \times \cdots \times B_m$ *are type-n.f.'s and* $M \equiv \langle M_1, \cdots, M_n \rangle$, $N \equiv \langle N_1, \cdots, N_m \rangle$ *are terms in* $\lambda^1\beta\eta\pi$ *such that*

$$x_1 : B_1, \cdots, x_m : B_m \vdash M_1, \cdots, M_n \quad M_i[\vec{N}/\vec{x}] =_{\beta\eta} y_i, \text{ for } 1 \leq i \leq n$$

$$y_1 : A_1, \cdots, y_n : A_n \vdash N_1, \cdots, N_m \quad N_j[\vec{M}/\vec{y}] =_{\beta\eta} x_j, \text{ for } 1 \leq i \leq m.$$

Then $n = m$ *and there exist permutations* σ, π *over* n *(and terms* P_i, Q_j *) such that*

$$M_i = \lambda \vec{u_i}.x_{\sigma_i} \vec{P}_i \qquad and \qquad N_j = \lambda \vec{v_j}.x_{\text{p}_i} \vec{Q}_j$$

Proof. By Lemma 4.2.6 (recall that we may assume that each M_i and N_j is in normal form) one has that M_i and N_j are in $\lambda^1\beta\eta$. Then,

$$M_i = \lambda \vec{u}_i.s_i \vec{P}_i \qquad and \qquad N_j = \lambda \vec{v}_j.t_j \vec{Q}_j$$

Note that s_i is a free variable (namely some x_j), since $M_i[\overrightarrow{N}/\overrightarrow{x}] =_{\beta\eta} y_i$. Indeed, if s_i is bound then M_i is $\lambda u_1 \cdots s_i \cdots u_k.s_i \overrightarrow{P}_i$ and $M_i[\overrightarrow{N}/\overrightarrow{x}]$ is $\lambda u_1 \cdots s_i \cdots u_k.s_i \overrightarrow{P}_i[\overrightarrow{N}/\overrightarrow{x}]$ so that s_i would still be a bound head variable and that there would be no way to reduce it to a term without abstraction. Similarly t_j is some y_i.

So there are two functions $\sigma : n \to m$, $\pi: m \to n$ such that

$$M_i = \lambda \overrightarrow{u}_i.x_{\sigma(i)} \overrightarrow{P}_i \text{ for } 1 \leq i \leq n, \qquad N_j = \lambda \overrightarrow{v}_j.y_{\pi(j)} \overrightarrow{Q}_j \text{ for } 1 \leq i \leq m.$$

In conclusion, for $1 \leq i \leq n$ we obtain:

$$
\begin{aligned}
y_i \;=_{\beta\eta}\; & M_i[\overrightarrow{N}/\overrightarrow{x}] \\
=_{\beta\eta}\; & (\lambda \overrightarrow{u_i}.x_{\sigma(i)} \overrightarrow{P}_i)[\overrightarrow{N}/\overrightarrow{x}] \\
=_{\beta\eta}\; & \lambda \overrightarrow{u_i}.N_{\sigma(i)}\{\overrightarrow{P}_i[\overrightarrow{N}/\overrightarrow{x}]\} \\
=_{\beta\eta}\; & \lambda \overrightarrow{u_i}.(\lambda \overrightarrow{v}_{\sigma(i)}.y_{\pi(\sigma(i))} \overrightarrow{Q}_{\sigma(i)})\{\overrightarrow{P}_i[\overrightarrow{N}/\overrightarrow{x}]\} \\
=_{\beta\eta}\; & \text{if } \overrightarrow{v}_{\sigma(i)} \text{ is longer than } \overrightarrow{P}_i \\
& \text{then } \lambda \overrightarrow{u_i}.\overrightarrow{v'}_{\sigma(i)}.y_{\pi(\sigma(i))} \overrightarrow{Q}_{\sigma(i)}[(\overrightarrow{P}_i[\overrightarrow{N}/\overrightarrow{x}])/\overrightarrow{u_i}]; \\
& \text{otherwise } \lambda \overrightarrow{u_i}.y_{\pi(\sigma(i))}\{\overrightarrow{Q}_{\sigma(i)}[(\overrightarrow{P}_i[\overrightarrow{N}/\overrightarrow{x}])/\overrightarrow{u_i}]\}\{\overrightarrow{P'}_i[\overrightarrow{N}/\overrightarrow{x}]\}
\end{aligned}
$$

We used here the accolades { and } to denote the scope of the substitutions.

In either case of the last equality, each term can reduce to y_i iff $y_i = y_{\pi(\sigma(i))}$ and each of the remaining Q's and P's orderly reduce to one of the bound variables, so that one can apply η, several times, at the end. The same holds for $N_j[\overrightarrow{M}/\overrightarrow{y}]$ for $1 \leq j \leq m$.

Thus $i = \pi(\sigma(i))$ for $1 \leq i \leq n$, and $j = \sigma(\pi(j))$ for $1 \leq j \leq m$ and we can conclude that $m = n$, σ is a permutation and π is its inverse. $\square$

We are then reduced to examining componentwise the terms which prove an isomorphism. The next point is to show that each component, indeed a term of $\lambda^1 \beta\eta$ by Lemma 4.2.6, yields an isomorphism.

4.3   Using finite hereditary permutations

In order to prove that the isomorphism between two type-n.f.'s can be expressed componentwise, we use the theorem from [Dez76] that we already discussed in the Introduction (there Theorem 1.9.1). The same result will also be applied to obtain, at last, the proof of our completeness theorem.

The first application of 1.9.1 we need is the following.

Proposition 4.3.1 *Let $M_1, \cdots, M_n$ and $N_1, \cdots, N_n$ and permutation σ satisfy all the assumptions in Lemma 4.2.8. Then $\lambda x_{\sigma(i)}.M_i{:}B_{\sigma(i)} \to A_i$ and $\lambda y_i.N_{\sigma(i)}{:}A_i \to B_{\sigma(i)}$ are invertible terms.*

Proof. For a suitable typing of the variables it is possible to build the following terms of $\lambda^1 \beta \eta$:

$$M = \lambda z.\lambda x_1 \cdots x_n.z M_1 \cdots M_n, \qquad N = \lambda z.\lambda y_1 \cdots y_n.z N_1 \cdots N_n.$$

It is an easy computation to check, by the definition of the M_i's and of the N_i's, that M and N are invertible. Moreover, they are (by the construction given in the Appendix) in normal form thus, by Dezani's theorem, (the erasures of) M and N are f.h.p.'s. This is enough to show that every M_i has only one occurrence of the x_i's (namely $x_{\sigma(i)}$), and similarly for the N_i's.

Thus we obtain $M_i[\vec{N}/\vec{x}] \equiv M_i[N_{\sigma(i)}/x_{\sigma(i)}] =_{\beta\eta} y_i$, for $1 \leq i \leq n$, and $N_i[\vec{M}/\vec{y}] \equiv N_i[M_{\pi(i)}/y_{\pi(i)}] =_{\beta\eta} x_i$, for $1 \leq i \leq n$,

Hence for each i, $\lambda x_{\sigma(i)}.M_i{:}B_{\sigma(i)} \to A_i$ and $\lambda y_i.N_{\sigma(i)}{:}A_i \to B_{\sigma(i)}$ are invertible. $\square$

As a result of all the work done so far, we can then focus on invertible terms whose types contain only " $\to$ ", that is, investigate componentwise the isomorphisms of type-n.f.'s. Of course, these isomorphisms will be given just by a fragment of the theory $Th^1_{\times T}$.

Definition 4.3.2 *Let* **Swap** *be the subtheory of $Th^1_{\times T}$ given by just the following proper axiom (plus the usual axioms and rules for "$=$"),*

$$\text{(\textbf{Swap})} \qquad A \to B \to C = B \to A \to C.$$

Swap is a subtheory of $Th^1_{\times T}$ by axioms 1 and 3 of $Th^1_{\times T}$.

Proposition 4.3.3 *Let A, B be type expressions with no occurrences of* **T** *or* $\times$. *Then $A \cong_d B \Rightarrow$* **Swap** $\vdash A = B$.

Proof. Suppose $A \cong_d B$ via M and N. As usual, we may assume without loss of generality that M and N are in normal form. By Lemma 4.2.6, M and N actually live in $\lambda^1 \beta \eta$, and by Theorem 1.9.5, $e(M)$ and $e(N)$ are f.h.p.'s. We prove **Swap** $\vdash A = B$ by induction on the depth of the Böhm-tree of M.

Depth 1: $M \equiv \lambda z : C.\ z$. Thus $M : C \to C$. Now, **Swap** $\vdash C = C$ by reflexivity.

Depth n+1: $M \equiv \lambda z : E.\ \lambda \vec{x}{:}\vec{D}.\ z\vec{N}_\sigma$.

Recall $z\vec{N}_\sigma = (\cdots (zN_{\sigma(1)}) \cdots N_{\sigma(n)})$, where if the ith abstraction in $\lambda \vec{x}{:}\vec{D}$ is $\lambda x_i{:}D_i$ then the erasure of $\lambda x_i{:}D_i.N_i$ is a f.h.p. Let F_i be the type of N_i. In order to type-check, we must have $E = (F_{\sigma(1)} \to \cdots \to F_{\sigma(n)} \to B)$ for some B. Thus the type of M is

$$(F_{\sigma(1)} \to \cdots \to F_{\sigma(n)} \to B) \to (D_{\sigma(1)} \to \cdots \to D_{\sigma(n)} \to B).$$

Since $\lambda x_i{:}D_i.N_i$ is a f.h.p, $\lambda x_i{:}D_i.N_i$ gives (half of) a provable isomorphism from D_i to F_i. By induction, since the height of the Böhm tree of (of the erasure of) each $\lambda x_i{:}D_i.N_i$ is less than the height of the Böhm tree of M, one has **Swap** $\vdash D_i = F_i$ for $1 \leq i \leq n$. By repeated use of the rules for "=", we get

Swap $\vdash (F_{\sigma(1)} \to \cdots \to F_{\sigma(n)} \to B) = (D_{\sigma(1)} \to \cdots \to D_{\sigma(n)} \to B)$

Hence it suffices to show

Swap $\vdash (D_{\sigma(1)} \to \cdots \to D_{\sigma(n)} \to B) = (D_1 \to \cdots \to D_n \to B)$

This is quite simple to show by repeated use of axiom (**Swap**) above in conjunction with the rules for equality.

□

Clearly, also the converse of Proposition 4.3.3 holds, since the "⇐" part in 4.3.3 is provable by a fragment of the proof in Theorem 1.9.9. Thus one has:

Swap $\vdash A = B \Longleftrightarrow A \cong_d B$ by terms in $\lambda^1\beta\eta$.

The result we aim at is just the extension of this fact to $Th^1_{\times T}$ and $\lambda^1\beta\eta\pi*$.

Theorem 4.3.4 (completeness) $B \cong_d A \Rightarrow Th^1_{\times T} \vdash B = A$

Proof. By Proposition 4.1.4, this is equivalent to proving $\mathrm{nf}(B) \cong_d \mathrm{nf}(A)$
$\Rightarrow Th^1_{\times T} \vdash \mathrm{nf}(B) = \mathrm{nf}(A)$.

Now, for $\mathrm{nf}(B) \equiv B_1 \times \cdots \times B_n$ and $A_1 \times \cdots \times A_m \equiv \mathrm{nf}(A)$, we have
shown, in Lemmas 4.2.7, 4.2.8 and Proposition 4.3.1, that $\mathrm{nf}(B) \cong_d \mathrm{nf}(A)$
$\Rightarrow n = m$ and there exist $M_1, \cdots, M_n$, $N_1, \cdots, N_n$ and a permutation σ such
that $\lambda x_{\sigma(i)}.M_i{:}B_{\sigma(i)} \to A_i$ and $\lambda y_i.N_{\sigma(i)}{:}B_i \to A_{\sigma(i)}$.

By 4.3.1, these terms are invertible too, for each i. Thus, by 4.3.3, **Swap**
$\vdash A_i = B_{\sigma(i)}$ and, hence, by the rules, $Th^1_{\times T} \vdash B = A$. $\square$

This proves completeness of $Th^1_{\times T}$ for $\lambda^1 \beta \eta \pi *$.

Together with the soundness result of Theorem 1.9.9, this gives us the
main result of this chapter.

Corollary 4.3.5 *The theory $Th^1_{\times T}$ exactly characterizes the isomorphic
types of $\lambda^1 \beta \eta \pi *$.*

Furthermore, equality is decidable in this theory, as we see as follows.

Corollary 4.3.6 *Given types A and B, it is decidable whether they are
isomorphic in all models of $\lambda^1 \beta \eta \pi *$. (And thus whether A and B name
isomorphic objects in all ccc's.)*

Proof. Let the type-n.f. of A be $A_1 \times \cdots \times A_n$ and that of B be
$B_1 \times \cdots \times B_n$ where neither the A_i's nor the B_j's contain any occurrences
of **T** or $\times$. (If one of A or B is **T**, the other must be as well if they are to
be isomorphic.) By Propositions 4.2.7 and 4.2.8, and Theorem 4.3.4, A and
B are isomorphic iff $m = n$ and there is a permutation σ over n such that
for $1 \leq i \leq n$, $A_i \cong_d B_{\sigma(i)}$. By Proposition 4.3.3, we know that **Swap** $\vdash A_i$
$= B_{\sigma(i)}$. Note that the axioms and rules of **Swap** do not change the length
of type expressions. Hence if **Swap** $\vdash A_i = B_{\sigma(i)}$, then A_i and $B_{\sigma(i)}$ have
the same length. We provide a decision procedure to determine if **Swap** $\vdash$
$A = B$ (and hence whether they are isomorphic in all models) by induction
on the length of A (and hence B). We restrict ourselves to type expressions
of the same length since otherwise they are not provably equal. If A and B
are both type symbols then they are equal if and only if they are the same
symbol. Suppose we have a decision procedure for all types of length less
than n, and A and B have length n. Decompose A and B into terms of
the form $A_1 \to \cdots \to A_k$ and $B_1 \to \cdots \to B_m$ where A_k and B_m are type
symbols. If A_k and B_m are different or $k \neq m$, then it is not the case that
Swap $\vdash A = B$. Otherwise, for each A_i determine if there is a distinct B_j
such that $\vdash A_k = B_m$. Each of these tests is decidable by hypothesis. If
each A_i can be paired with a distinct B_j, then **Swap** $\vdash B = A$. Otherwise
it fails. The proof of the correctness of this decision proceed follows the
same lines as the proof of the ($\Rightarrow$) direction of Proposition 4.3.3. $\square$

4.4   The complete theories of $\lambda^1 \beta \eta \pi$ and $\lambda^1 \beta \eta *$

The method we used in this chapter to prove the completeness of $Th^1_{\times \mathbf{T}}$ for $\lambda^1 \beta \eta \pi *$ actually provides a very modular separation of the additional type constructor $\times$ and $\mathbf{T}$. Indeed, in order to reduce completeness of $Th^1_{\times \mathbf{T}}$ for $\lambda^1 \beta \eta \pi *$ to completeness of Th^1 for $\lambda^1 \beta \eta$, we first got rid of the terminal object, using the type rewriting system of Section 4.1, which at the same time gives us a usable type normal form. Then we showed, in Sections 4.2 and 4.3, that such type normal forms (no longer containing occurrences of $\mathbf{T}$) are isomorphic if and only if their components are (up to a permutation) isomorphic componentwise; furthermore, this componentwise isomorphism is provable using just the axiom **Swap**, which is derivable in $Th^1_{\times \mathbf{T}}$.

We can see than that the seven axioms making up our theory $Th^1_{\times \mathbf{T}}$ are used as follows along the proofs:

- Reduction from $\lambda^1 \beta \eta \pi *$ to $\lambda^1 \beta \eta \pi$ (via type rewriting): use of all the axioms involving terminal object ($A \times \mathbf{T} = A$, $A \to \mathbf{T} = \mathbf{T}$ and $\mathbf{T} \to A = A$), currification and distribution of product w.r.t. the arrow (axioms $(A \times B) \to C = A \to (B \to C)$ and $A \to (B \times C) = (A \to B) \times (A \to C)$), and associativity.

- Reduction from $\lambda^1 \beta \eta \pi$ to $\lambda^1 \beta \eta$ (Section 4.2, then Proposition 4.3.1): associativity and commutativity of product (axioms $A \times B = B \times A$ and $A \times (B \times C) = (A \times B) \times C$)

- Isomorphism in $\lambda^1 \beta \eta$: use of axiom **Swap** only (derivable from commutativity and currying)

But this very procedure gives us more than just completeness of $Th^1_{\times \mathbf{T}}$. Indeed, if in the language there is no $\mathbf{T}$ type, then the first step does not use axioms involving $\mathbf{T}$, and the second reduction step can be performed entirely inside $\lambda^1 \beta \eta \pi$, as we already noticed along the way in Proposition 4.2.3. This tells us that the complete theory of isomorphisms for $\lambda^1 \beta \eta \pi$ is given by the following:

Definition 4.4.1 $Th^1_\times$, the theory of isomorphic types of $\lambda^1 \beta \eta \pi$, is a theory of equality plus the following axiom schemas:

1. $A \times B = B \times A$

2. $A \times (B \times C) = (A \times B) \times C$

3. $(A \times B) \to C = A \to (B \to C)$

4. $A \to (B \times C) = (A \to B) \times (A \to C)$

On the other side, if it is the product type-constructor that is omitted, then the first step is performed without any axiom related to the product; similarly the second step, where there is no need to look for permutations of components (as there is always just one component). Anyway, in the last step, axiom **Swap** is no longer derivable in the absence of axioms involving products, so one needs to add it explicitly, and we get the complete theory of isomorphisms for $\lambda^1 \beta\eta*$ as follows:

Definition 4.4.2 $Th^1_{\mathbf{T}}$, the theory of isomorphic types of $\lambda^1 \beta\eta*$, is a theory of equality plus the following axiom schemas:

1. $A \rightarrow (B \rightarrow C) = B \rightarrow (A \rightarrow C)$ (**Swap**)

2. $A \rightarrow \mathbf{T} = \mathbf{T}$

3. $\mathbf{T} \rightarrow A = A$

The very same schema will appear in the proofs involving the second-order systems. Indeed, our technique for dealing with the full systems $\lambda^1 \beta\eta\pi*$ and $\lambda^2 \beta\eta\pi*$ already automatically provides an answer for the subsystems.

Chapter 5

Second-Order Isomorphic Types

This chapter is dedicated to the proof of completeness of $Th_{\times\mathbf{T}}^2$ for iso-morphisms in $\lambda^2\beta\eta\pi*$. This proof is by far the most complex present in this book, because in the second-order case we have to face the problem of invertibility of terms almost anew, and we can no longer avoid it as we did for the first-order systems.

It is not an easy task to provide the reader with a presentation that is at the same time clear, easily readable and fully detailed, but we tried to do our best to increase readability without sacrificing the technical precision. To improve readability, the details of the more difficult and lengthy proofs are moved into an Appendix, and only sketched in the main body of the chapter. Finally, every section has been provided with an introductory paragraph to outline the work done inside.

5.1   Towards completeness

As we have seen in the survey, the only proof technique that we can hope to apply to the case of $\lambda^2\beta\eta\pi*$ is the one based on the syntactic charac-terization of invertible terms. Unfortunately, no such characterization is known up to now for $\lambda^2\beta\eta\pi*$, so we need to reduce our original problem to a simpler one. In this section we will show that we can actually restrict our attention to isomorphisms of a special form and to a particular class of invertible terms, for which we will later be able to provide a syntactic characterization.

5.1.1   Outline of the section

- **Reduction to a subclass of types.** We identify two relevant classes of types: types not containing products or $\mathbf{T}$, which we call *simple types*, and products of simple types, which we call *regular types*. We show that $Th_{\times\mathbf{T}}^2$ is complete for isomorphisms of types if and only if it is complete for isomorphisms between *regular type*.

- **Reduction to a subclass of terms.** We show that any isomorphism between *regular* types can be proved by invertible terms whose free variables have *simple* types (we call them *canonical* invertible terms).

- **Overall achievement of this section.** We reduce the problem of completeness of $Th_{\times\mathbf{T}}^2$ to the problem of completeness for isomor-phisms between *regular* types proved by *canonical* invertible terms, and for these we will be able to provide a syntactic characterization.

5.1.2 Reduction to a subclass of types

We introduce here a type-rewriting system, suggested by the form of the axioms of $Th^2_{\times\mathbf{T}}$, and the corresponding type normal form. We will then show that two types are isomorphic if and only if their normal forms are. It is to be noticed that the normal form is essentially the usual conjunctive normal form for (second-order) propositional calculus. The axioms of $Th^2_{\times\mathbf{T}}$ suggest the following rewrite system $\mathcal{R}^2$ for types (essentially $Th^2_{\times\mathbf{T}}$ without commutativity):

Definition 5.1.1 *(Type-rewriting $\mathcal{R}^2$) Let "$\mathcal{R}^2$" be the transitive and substitutive type-reduction relation generated by:*

1.	$A \times (B \times C)$	$>$	$(A \times B) \times C$		5.	$A \times \mathbf{T}$	$>$	A
2.	$(A \times B) \to C$	$>$	$A \to (B \to C)$		6.	$\mathbf{T} \times A$	$>$	A
3.	$A \to (B \times C)$	$>$	$(A \to B) \times (A \to C)$		7.	$A \to \mathbf{T}$	$>$	$\mathbf{T}$
4.	$\forall X.A \times B$	$>$	$\forall X.A \times \forall X.B$		8.	$\mathbf{T} \to A$	$>$	A
					9.	$\forall X.\mathbf{T}$	$>$	$\mathbf{T}$.

The system $\mathcal{R}^2$ yields an obvious notion of normal form for types (type-n.f.), that is, when no type reduction is applicable. Note that 5, 6, 8, *eliminate the* $\mathbf{T}$'s, while 3 and 4 *bring the* $\times$ *outside*. It is then easy to observe that each type-n.f. is $\mathbf{T}$ or has the structure $S_1 \times \cdots \times S_n$ where each S_i does not contain $\mathbf{T}$ or $\times$. We write nf(S) for the normal form of S (there is exactly one, see 5.1.2), and say that a normal form is nontrivial if it is not $\mathbf{T}$.

Proposition 5.1.2 *Each type has a unique type normal form in $\mathcal{R}^2$.*

Proof. Using the REVE system [Les83, Les86], this is straightforward, but in Appendix C we provide also a direct proof. □

Types in normal form have a very simple shape, that can be described as follows:

Definition 5.1.3 (simple types, regular types) *A type is* **simple** *when there is no occurrence of products or* $\mathbf{T}$*'s in it. A type is* **regular** *when it is either* $\mathbf{T}$ *or a finite product* $A_1 \times \cdots \times A_n$*, where the A_i are simple types.*

Remark 5.1.4 *Every type normal form is regular. Furthermore, whenever $A \cong_d B$, either nf(A) and nf(B) are both $\mathbf{T}$, or they are both not $\mathbf{T}$.*

Indeed, a nontrivial type-n.f. cannot be isomorphic to $\mathbf{T}$, as is easily seen by taking a nontrivial model, so the case $(A_1 \times \cdots \times A_n) \cong_d \mathbf{T}$ is not possible. Anyway, since all the work done in this section is purely syntactic, we give also an easy syntactic proof of this fact in Proposition C.1 in Appendix C. Now, $\mathcal{R}^2 \vdash A > B$ implies $Th^2_{\times \mathbf{T}} \vdash A = B$, and using the soundness of $Th^2_{\times \mathbf{T}}$ proved in 1.9.9, we get that any reduction $\mathcal{R}^2 \vdash A > B$ is witnessed (or *proved*, in the *types-as-propositions analogy*) by an invertible term $M : A \to B$. Moreover, one clearly has:

Corollary 5.1.5 $Th^2_{\times \mathbf{T}} \vdash A = nf(A)$ and, thus, $Th^2_{\times \mathbf{T}} \vdash A = B \iff Th^2_{\times \mathbf{T}} \vdash nf(A) = nf(B)$.

The same holds for $\cong_d$:

Proposition 5.1.6 $A \cong_d B \iff nf(A) \cong_d nf(B)$.

Proof. Recall that $A \cong_d B$ iff there is an invertible term $M : A \to B$, and $nf(A) \cong_d nf(B)$ iff there is an invertible term $M' : nf(A) \to nf(B)$. So it suffices to show that it is possible to turn invertible terms of type $A \to B$ into invertible terms of type $nf(A) \to nf(B)$, and vice-versa. Given types A and B, assume that $F : A \to nf(A)$ and $G : B \to nf(B)$ prove the reductions to type-n.f. Then $M : A \to B$ is invertible $\iff$ there exists an invertible term $M' : nf(A) \to nf(B)$, such that $M = G^{-1} \circ M' \circ F$.

($\Leftarrow$) Set $M^{-1} \equiv (G^{-1} \circ M' \circ F)^{-1} \equiv F^{-1} \circ M'^{-1} \circ G$, then M is invertible.

($\Rightarrow$) Just set $M' = G \circ M \circ F^{-1}$. Then $M'^{-1} \equiv F \circ M^{-1} \circ G^{-1}$ and M' is invertible. $\square$

The diagram in Figure 5.1 shows what's going on in the corollary.

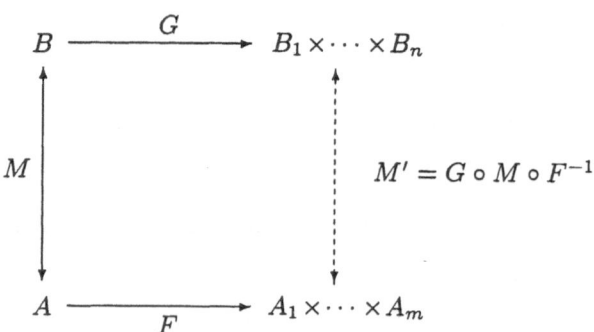

Figure 5.1: Reduction to a subclass of isomorphic types.

We get hence the main result in this subsection, that allows us to restrict the analysis to isomorphisms between nontrivial regular types.

Proposition 5.1.7 $Th^2_{\times \mathbf{T}}$ *is complete for* $\cong_d$ *iff* $Th^2_{\times \mathbf{T}}$ *is complete for* $\cong_d$ *restricted to regular types different from (and hence not containing)* **T**.

Proof. ($\Leftarrow$) If $A \cong_d B$, then by 5.1.4 either $nf(A) \equiv \mathbf{T} \equiv nf(B)$, or $nf(A) \equiv (A_1 \times \cdots \times A_n) \cong_d (B_1 \times \cdots \times B_m) \equiv nf(B)$, where no occurrence of **T** can appear in either type. In both cases the result follows by

$$Th^2_{\times \mathbf{T}} \vdash A = B \Longleftrightarrow Th^2_{\times \mathbf{T}} \vdash nf(A) = nf(B) \Longleftrightarrow A \cong_d B \Longleftrightarrow nf(A) \cong_d nf(B)$$

The first and third equivalences are just Corollary 5.1.5 and Proposition 5.1.6. The second equivalence is trivially satisfied in the case $nf(A) \equiv \mathbf{T} \equiv nf(B)$; otherwise both types are different from **T**, and then this equivalence becomes just the hypothesis of completeness for regular types different from **T**.

($\Rightarrow$) By definition, as regular types are types. $\square$

Thus we have shown that the characterization of definable isomorphisms between arbitrary types can be reduced to the characterization of definable isomorphisms in the class of **regular** types different from **T**.

5.1.3   Reduction to a subclass of terms

There is another simplification we can perform, though. Since equality on lambda terms is substitutive, it is easy to show that we can consider only invertible terms with free variables of **simple** types.

Proposition 5.1.8 *Let* $M{:}A \to B$ *be an invertible term with inverse* $N{:}B \to A$. *Then there exist invertible terms* $M'{:}A \to B$ *and* $N'{:}B \to A$ *whose free variables have simple types.*

Proof. For every free variable $x : A$ it is easy to build a term $t_x{:}A$ whose free variables have simple types. If A is simple, then $t_x = x$. Otherwise, consider the type normal form $nf(A)$ of A, and let $G_A{:}nf(A) \to A$ be a (closed) invertible term proving the isomorphism $A \cong_d nf(A)$ (as in 5.1.6). If $nf(A) = A_1 \times \cdots \times A_n$, with $n > 1$, then we can choose fresh variables $z_1{:}A_1, \cdots, z_n{:}A_n$ and put $t_x = G_A \langle z_1, \cdots, z_n \rangle{:}A$, since the A_i are simple and G_A is closed. If $nf(A) = A_1 \neq \mathbf{T}$ we can choose a fresh variable $z_1{:}A_1$ and put $t_x = (G_A\ z_1)$, since A_1 is simple and G_A is closed. Otherwise, $nf(A) = \mathbf{T}$ and we can put $t_x = (G_A*)$, which is closed. Now we can show that $M[\overrightarrow{t_x}/\overrightarrow{x}]$ and $N[\overrightarrow{t_x}/\overrightarrow{x}]$, where $\overrightarrow{x}$ are all the free variables in M and N, are the required invertible terms M' and N'. Indeed, the free variables in

$M[\overrightarrow{t_x}/\overrightarrow{x}]$ and $N[\overrightarrow{t_x}/\overrightarrow{x}]$ have simple types as they are included in the free variables of the t_x's. Furthermore, by substitutivity of the equality, $N \circ M = I_A$ implies $(N \circ M)[\overrightarrow{t_x}/\overrightarrow{x}] = I_A[\overrightarrow{t_x}/\overrightarrow{x}]$, which is $(N[\overrightarrow{t_x}/\overrightarrow{x}] \circ M[\overrightarrow{t_x}/\overrightarrow{x}]) = I_A$. Similarly, $(M[\overrightarrow{t_x}/\overrightarrow{x}] \circ N[\overrightarrow{t_x}/\overrightarrow{x}]) = I_B$. $\square$

Proposition 5.1.7 and 5.1.8 allow, without loss of generality, to restrict the analysis to invertible terms between regular types (different from $\mathbf{T}$). By Proposition 5.1.8, we can assume that the free variables in these invertible terms have simple types only. We will call these terms **canonical bijections**, or simply **canonical** terms.

Definition 5.1.9 (canonical bijections) *A term is a* **canonical bijection** *if it is an invertible term mapping regular types (different from $\mathbf{T}$) into regular types (different from $\mathbf{T}$) and if all its free variables have simple types.*

The next step will be to find a syntactic characterization for them.

5.2 Characterizing canonical terms: from $\lambda^2\beta\eta\pi*$ to $\lambda^2\beta\eta$

In this section we follow a very natural intuition: if two regular types $A_1 \times \cdots \times A_m$ and $B_1 \times \cdots \times B_n$ are isomorphic, we expect them to have the same number of components (i.e., $n = m$). Furthermore, we expect that such isomorphism can be decomposed into an n-tuple of independent, simpler isomorphisms between the different components of the regular types. These componentwise isomorphisms should not involve products or $\mathbf{T}$, and be invertible terms of $\lambda^2\beta\eta$, rather then of $\lambda^2\beta\eta\pi*$. Essentially, this allows to express canonical bijections of $\lambda^2\beta\eta\pi*$ in terms of invertible terms of $\lambda^2\beta\eta$, for which a syntactic characterization is known from [BL85].

5.2.1 Outline of the section

We define a notion of **coordinates** for a canonical bijection. Such coordinates will be the body of the seeked componentwise isomorphisms, and will allow to give a syntactic characterization of canonical bijections.

- **Projection of invertibility over coordinates.** We show that the invertibility property of a canonical bijection determines a similar property, which we call **distributed invertibility**, for its set of coordinates.

- **Reduction of coordinates to $\lambda^2\beta\eta$.** Using in an essential way this property of coordinates of a canonical bijection, we can show with a difficult syntactic proof that such coordinates are already terms of $\lambda^2\beta\eta$. This proof is intimately related to the method employed in the original characterization of invertible terms in pure lambda calculus [Dez76].

- **Syntactic Characterization of canonical bijections.** Once we know that coordinates live in $\lambda^2\beta\eta$, we can use the syntactic characterization of invertible terms in $\lambda^2\beta\eta$ provided in [BL85] to obtain a simple syntactic characterization of canonical bijections.

- **Overall achievement of this section.** We explicitly describe the syntactic shape of a canonical bijection. This result will allow to show completeness of $Th^2_{\times\mathbf{T}}$ by structural induction on the canonical bijections and also provides us with a decidable test of invertibility for all terms in $\lambda^2\beta\eta\pi*$.

5.2.2 Projection of invertibility over coordinates

A canonical bijection maps a finite product $A_1\times\cdots\times A_m$ of nonproduct types into another finite product $B_1\times\cdots\times B_n$ of nonproduct types. It is natural to study the behavior of the bijection on each one of the components B_i of the arrival type separately. To do so, we introduce a notion of coordinate as follows:

Definition 5.2.1 (coordinates) *For a canonical bijection*

$$M : A_1\times\cdots\times A_m \to B_1\times\cdots\times B_n,$$

where $A_1\times\cdots\times A_m$ and $B_1\times\cdots\times B_n$ are regular types, define the collection of its coordinates *as the sequence $\overrightarrow{M} = [M_1, \cdots, M_n]$, where M_i is n.f.$(\mathrm{p}_i(M\langle x_1,\cdots,x_m\rangle))$ and $x_1,\cdots,x_m$ are fresh variables.*

Here $\mathrm{p}_i M$ stands for the ith projection of the term M, which is more precisely written as $p_i^n M$ using the notation A.1 of the Appendix. Anyway, since n is clear from the context, we will often omit it.

Remark 5.2.2 *The type of a coordinate and of the free variables of a coordinate do not contain products or $\mathbf{T}$ types.*

A canonical bijection can be represented by its coordinates.

Proposition 5.2.3 *Every canonical bijection*

$$M : A_1 \times \cdots \times A_m \to B_1 \times \cdots \times B_n$$

can be written as $\lambda z.(\lambda x_1 \cdots x_m.(\langle M_1, \cdots, M_n \rangle))(\mathrm{p}_1 z)\cdots(\mathrm{p}_m z)$ *in terms of its coordinates* M_i's.

Proof. Due to surjective pairing, we have

$M \langle x_1, \cdots, x_m \rangle$

$\quad =_{\beta^2 \eta^2 \pi *} \quad \langle \mathrm{p}_1(M \langle x_1, \cdots, x_m \rangle), \cdots, \mathrm{p}_n(M \langle x_1, \cdots, x_m \rangle) \rangle$

$\quad =_{\beta^2 \eta^2 \pi *} \quad \langle n.f.(\mathrm{p}_1(M \langle x_1, \cdots, x_m \rangle)), \cdots, n.f.(\mathrm{p}_n(M \langle x_1, \cdots, x_m \rangle)) \rangle$

$\quad =_{\beta^2 \eta^2 \pi *} \quad \langle M_1, \cdots, M_n \rangle.$

Notice that, by standard currying,

$$M =_{\beta^2 \eta^2 \pi *} \lambda z.(\lambda x_1 \cdots x_m.(M \langle x_1, \cdots, x_m \rangle))(\mathrm{p}_1 z)\cdots(\mathrm{p}_m z)$$

so we finally get $M = \lambda z.(\lambda x_1 \cdots x_m.(\langle M_1, \cdots, M_n \rangle))(\mathrm{p}_1 z)\cdots(\mathrm{p}_m z).$ $\square$

The property of invertibility of the original bijection is reflected in a similar property for the collection of its coordinates (point 1 of the following proposition).

Proposition 5.2.4 (Properties of Coordinates)
Let $M : A_1 \times \cdots \times A_m \to B_1 \times \cdots \times B_n$, $N : B_1 \times \cdots \times B_n \to A_1 \times \cdots \times A_m$
be canonical bijections in $\lambda^2 \beta \eta \pi *$. *Then there exist sets of coordinates* $\overrightarrow{N}$
$= [N_1, \cdots, N_m]$ *and* $\overrightarrow{M} = [M_1, \cdots, M_n]$ *with* $\overrightarrow{x} = \{x_1 : A_1, \cdots, x_m : A_m\} \subseteq$
$FV(\overrightarrow{M})$ *and* $\overrightarrow{y} = \{y_1 : B_1, \cdots, y_n : B_n\} \subseteq FV(\overrightarrow{N})$ *s.t. no* x_i *is free in any*
N_j, *no* y_i *is free in any* M_j, *and*

1. $M_i[\overrightarrow{N}/\overrightarrow{x}] =_{\beta^2 \eta^2 \pi *} y_i$ *and* $N_i[\overrightarrow{M}/\overrightarrow{y}] =_{\beta^2 \eta^2 \pi *} x_j$

2. $FV(\overrightarrow{M}) \vdash M_i : B_i$ *and* $FV(\overrightarrow{N}) \vdash N_j : A_j$

3. *no type expression occurring in* $FV(\overrightarrow{M}) \cup FV(\overrightarrow{N})$ *contains occurrences of* $\times$ *or* $\mathbf{T}$.

Proof. The condition about the variables ($\overrightarrow{x}$ and $\overrightarrow{y}$) is trivially satisfied by a suitable choice of the new x_i's and y_i's in the coordinates.

(1) By Proposition 5.2.3 we have

$$M =_{\beta^2 \eta^2 \pi *} \lambda z.(\lambda x_1 \cdots x_m.M')(\mathrm{p}_1 z)\cdots(\mathrm{p}_m z)$$

$$N =_{\beta^2 \eta^2 \pi *} \lambda z.(\lambda y_1 \cdots y_n.N')(\mathrm{p}_1 z)\cdots(\mathrm{p}_n z)$$

where

$$M' : B_1 \times \cdots \times B_n =_{\beta^2 \eta^2 \pi *} \langle M_1, \cdots, M_n \rangle$$

and

$$N' : A_1 \times \cdots \times A_m =_{\beta^2 \eta^2 \pi *} \langle N_1, \cdots, N_m \rangle$$

so that $M \circ N =_{\beta^2 \eta^2 \pi *} \lambda w.w$ gives us the following equalities (we drop the suffix of $=$):

$$
\begin{aligned}
M \circ N \;=\;& \lambda w.(M(Nw)) \\
=\;& \lambda w.(\lambda z.(\lambda x_1 \cdots x_m.M')(\mathrm{p}_1 z)\cdots(\mathrm{p}_m z))(Nw)) \\
=\;& \lambda w.(\lambda x_1 \cdots x_m.M')(\mathrm{p}_1(Nw))\cdots(\mathrm{p}_m(Nw)) \\
=\;& \lambda w.M'[\mathrm{p}_1(Nw)\cdots \mathrm{p}_m(Nw)/x_1 \cdots x_m] \\
\equiv\;& \lambda w.M'[\overrightarrow{\mathrm{p}_i(Nw)}/\overrightarrow{x}] \\
=\;& \lambda w.\langle \mathrm{p}_1 M', \cdots, \mathrm{p}_n M' \rangle [\overrightarrow{\mathrm{p}_i(Nw)}/\overrightarrow{x}] \\
=\;& \lambda w.\langle M_1[\overrightarrow{\mathrm{p}_i(Nw)}/\overrightarrow{x}], \cdots, M_n[\overrightarrow{\mathrm{p}_i(Nw)}/\overrightarrow{x}] \rangle \\
& \text{as } M_i \equiv n.f.(\mathrm{p}_i M') \\
=\;& \lambda w.\langle M_1[(\overrightarrow{N_i[\overrightarrow{\mathrm{p}_j w}/\overrightarrow{y}]})/\overrightarrow{x}], \cdots, M_n[(\overrightarrow{N_i[\overrightarrow{\mathrm{p}_j w}/\overrightarrow{y}]})/\overrightarrow{x}] \rangle
\end{aligned}
$$

$$
\begin{aligned}
\text{as } Nw \;=\;& (\lambda z.(\lambda y_1 \cdots y_n.N')(\mathrm{p}_1 z)\cdots(\mathrm{p}_n z))w \\
=\;& (\lambda y_1 \cdots y_n.N')(\mathrm{p}_1 w)\cdots(\mathrm{p}_n w) \\
=\;& N'[\overrightarrow{\mathrm{p}_j w}/\overrightarrow{y}] \\
=\;& \langle \mathrm{p}_1 N', \cdots, \mathrm{p}_m N' \rangle [\overrightarrow{\mathrm{p}_j w}/\overrightarrow{y}] \\
=\;& \langle \mathrm{p}_1 N'[\overrightarrow{\mathrm{p}_j w}/\overrightarrow{y}], \cdots, \mathrm{p}_m N'[\overrightarrow{\mathrm{p}_j w}/\overrightarrow{y}] \rangle \\
=\;& \langle N_1[\overrightarrow{\mathrm{p}_j w}/\overrightarrow{y}], \cdots, N_m[\overrightarrow{\mathrm{p}_j w}/\overrightarrow{y}] \rangle,
\end{aligned}
$$

where for substitutions like $[\mathrm{p}_1(Nw), \cdots, \mathrm{p}_m(Nw)/x_1, \cdots, x_m]$ we have used the compact notation $[\overrightarrow{\mathrm{p}_i(Nw)}/\overrightarrow{x}]$.

But we also have that

$$
\begin{aligned}
M \circ N \;&=_{\beta^2 \eta^2 \pi *} \; \lambda w.w \text{ (by hypothesis)} \\
&=_{\beta^2 \eta^2 \pi *} \; \lambda w.\langle \mathrm{p}_1 w, \cdots, \mathrm{p}_n w \rangle \text{ (by surjective pairing).}
\end{aligned}
$$

By transitivity, then

$$\lambda w.\langle \mathrm{p}_1 w, \cdots, \mathrm{p}_n w \rangle =_{\beta^2 \eta^2 \pi *} \lambda w.\langle N_1[\overrightarrow{\mathrm{p}_j w}/\overrightarrow{y}], \cdots, N_m[\overrightarrow{\mathrm{p}_j w}/\overrightarrow{y}] \rangle.$$

From this equality it is easy to derive (by applications and projections) that

$$M_k[\overrightarrow{(N_i[\overline{p_j w}/\overrightarrow{y}])}/\overrightarrow{x}] =_{\beta^2 \eta^2 \pi_*} p_k w.$$

and then the Church-Rosser Property of the rewrite system for our calculus gives us

$$M_k[\overrightarrow{(N_i[\overline{p_j w}/\overrightarrow{y}])}/\overrightarrow{x}] \overset{\beta^2 \eta^2 \pi_*}{\longrightarrow} p_k w.$$

By substitution properties

$$M_k[\overrightarrow{(N_i[\overline{p_j w}/\overrightarrow{y}])}/\overrightarrow{x}] \equiv M_k[\overrightarrow{N}/\overrightarrow{x}][\overline{p_j w}/\overrightarrow{y}],$$

since no y_j is free in M, so

$$M_k[\overrightarrow{N}/\overrightarrow{x}][\overline{p_j w}/\overrightarrow{y}] \overset{\beta^2 \eta^2 \pi_*}{\longrightarrow} p_k w$$

Now, by Theorem A.4 (which is to be found in the Appendix) applied to M_k and w, we can deduce that $M_k[\overrightarrow{N}/\overrightarrow{x}] \overset{\beta^2 \eta^2 \pi_*}{\longrightarrow} y_k$, and hence $M_k[\overrightarrow{N}/\overrightarrow{x}] = y_k$, as required.

Analogously for $N \circ M$.

(2) Follows from the construction of the M_i and N_j.

(3) Follows from the definition of regular types and the hypothesis on the free variables of M and N.

□

So, to every bijection we can associate a set of coordinates that not only represents them in the sense of Proposition 5.2.3 but also enjoys a kind of **distributed invertibility** property (the one expressed by 2 in Proposition 5.2.4 above). This property, together with the other side conditions in the statement of Proposition 5.2.4, will be used in a crucial way to prove that coordinates are terms of $\lambda^2 \beta \eta$. For this reason, we extend it to sequences containing not only terms but also types, and we give a name to it.

Definition 5.2.5 (Distributed Invertibility)

Given $\overrightarrow{N} = [N_1, \cdots, N_m]$ and $\overrightarrow{M} = [M_1, \cdots, M_n]$ sequences of second-order λ-terms in normal form *and/or second-order type variables*, choose $\overrightarrow{x} = V_1, \cdots, V_m \subseteq FV(\overrightarrow{M})$ and $\overrightarrow{y} = W_1, \cdots, W_n \subseteq FV(\overrightarrow{N})$ sequences of variables, which can be either type variables X_i which are part of the list $\overrightarrow{M} = [M_1, \cdots, M_n]$ p(res. type variables Y_j from $\overrightarrow{N} = [N_1, \cdots, N_m]$), or term variables $x_i : A_i$ (resp. $y_i : B_i$). We require that $\overrightarrow{x} \cap FV(\overrightarrow{N}) = \emptyset$ and $\overrightarrow{y} \cap FV(\overrightarrow{M}) = \emptyset$.

We will write $DistInv(\overrightarrow{M}, \overrightarrow{x}, \overrightarrow{N}, \overrightarrow{y})$ if the following conditions hold:

1. no type expression occurring in $FV(\overrightarrow{M}) \cup FV(\overrightarrow{N})$ contains occurrences of $\times$ or $\mathbf{T}$,

2. $FV(\overrightarrow{M}) \vdash M_i : B_i$ if M_i is a term and $FV(\overrightarrow{N}) \vdash N_j : A_j$ if N_j is a term,

3. $M_i[\overrightarrow{N}/\overrightarrow{x}] = W_i$ and $N_j[\overrightarrow{M}/\overrightarrow{y}] = V_j$.

Remark 5.2.6 *Notice that this last equality is on terms or types, depending on the nature of the objects that are equated. In particular, this implies that M_i and W_i (resp. N_j and V_j) are either both types or both terms. Furthermore, the M_i (resp. N_j) that are types must be simple type variables; otherwise they could not be equal to the type variables W_i (resp. V_j), regardeless of the substitution we apply to them.*

Intermezzo

The coordinates of a canonical bijection M and those of a canonical bijection N that is its inverse are in the same number (i.e., $m = n$) and they are actually terms of the pure calculus $\lambda^2 \beta \eta$, as we will show in the next Subsection. The proof of such a fact, though, is far from being a simple one. To understand better the reason of this complexity, and the technique we will use to overcome it, it is convenient here to recall the proof techniques used in previous works. In the previous chapter, exactly the same proof strategy is used up to this point but for the limited case of the calculus $\lambda^1 \beta \eta \pi *$, which has no second-order features. There Proposition 4.2.7 has the same flavor as Proposition 5.2.4 here, but there it is also possible to show, by induction, that the coordinates are terms of the simple typed lambda calculus. The proof uses in an essential way the fact that the types of a term of $\lambda^1 \beta \eta \pi *$ and its free variables carry enough information to exclude the presence of products or terminal constants in the coordinates. This is due to the following relevant facts.

- *Every term in n.f. of $\lambda^1 \beta \eta \pi *$, whose type contains no occurrence of* **T**, *has no occurrence of* *A *constants.*

  Proof. This is Lemma 4.2.2. $\square$

- *Terms in n.f. of $\lambda^1 \beta \eta \pi$, whose type is arrow-only, belong to $\lambda^1 \beta \eta$.*

  Proof. This is Lemma 4.2.6. $\square$

Once this reduction from $\lambda^1 \beta \eta \pi *$ to $\lambda^1 \beta \eta$ is done, and it is then possible to prove easily that the coordinates of a canonical bijection M and those of a canonical bijection N that is its inverse are in the same number (i.e., $m = n$), as it is done there in Lemma 3.8.

Unfortunately, these statements do not hold any longer when we consider second-order terms, even in normal form, as the following example shows.

Example 5.2.7 Let A and B be *simple* types and consider the term

$$x : \forall X.(X \to B), y : A \vdash x[A \times \mathbf{T}]\langle y, * \rangle : B.$$

The term $x[A \times \mathbf{T}]\langle y, * \rangle$ is in normal form and no product or **T** appear in its type or the type of any of its variables, but it contains an occurrence of **T**, a product type, * and a pair as subterms. $\square$

Actually, these relevant properties do hold for those second-order terms that are coordinates of invertible terms, but we can no longer provide two separate proofs: one for reducing to the pure calculus $\lambda^2 \beta \eta$ and one to show that $m = n$. We must show these properties at the same time in a complex inductive proof that needs also several additional invariants, including the property *DistInv*.

5.2.3   Reduction of coordinates to $\lambda^2 \beta \eta$

To show that the coordinates of an invertible term of $\lambda^2 \beta \eta \pi *$ are indeed terms of the simpler calculus $\lambda^2 \beta \eta$, we will need to prove a main lemma that is comparable for its complexity to the original characterization of invertible terms of the pure λ-calculus provided by Dezani [Bar84, Dez76]. Actually, unless the reader is familiar with the proof technique by Dezani, even the statement of the lemma can be unreadable without a proper explanation. For this reason, in this section we will focus on showing how the statement of such a complex lemma arises, rather than on its proof, which does not present overwhelming technical difficulties and is therefore deferred to Appendix B.

Towards a better lemma

As we have seen in the Intermezzo, in the case of the calculus $\lambda^2\beta\eta\pi*$, we cannot prove independently that the number of coordinates of an invertible term M is the same as the number of coordinates of any inverse N *and* that such coordinates are terms of the simpler calculus $\lambda^2\beta\eta$. Actually, it turns out that this last property needs a strengthened version of the first one.

To start, let's recall how in Lemma 4.2.8 it is proved that the length of the coordinates of two invertible terms M and N of $\lambda^1\beta\eta\pi*$ that are each other's inverses is the same (we rephrase the statement using the terminology of this paper).

Lemma 4.2.8 (also Lemma 3.8 of [BDCL92]) *Let* $A_1 \times \cdots \times A_n$ *and* $B_1 \times \cdots \times B_m$ *be type normal forms and* $M_1 \cdots M_n$, $N_1, \cdots, N_n$ *be coordinates of invertible terms* M *and* N *of* $\lambda^1\beta\eta\pi*$. *Then*

- $n = m$

- *there exist permutations* σ, π *over* n *(and terms* P_i, Q_j*) such that*

$$M_i = \lambda \overrightarrow{u_i}.x_{\sigma_i} \overrightarrow{P}_i \qquad and \qquad N_j = \lambda \overrightarrow{v_j}.y_{\mathrm{p}_j} \overrightarrow{Q}_j$$

(where the x_{σ_i} *and* y_{p_i} *are free in* M_i *and* N_j, *respectively).*

Proof. First, one already knows that the coordinates M_i and N_j, which are terms in normal form, are terms of the simpler calculus $\lambda^1\beta\eta$, so that

$$M_i = \lambda \overrightarrow{u}_i.s_i \overrightarrow{P}_i \qquad and \qquad N_j = \lambda \overrightarrow{v}_j.t_j \overrightarrow{Q}_j.$$

Then it is easy to notice that s_i is a free variable (namely some x_j), since $M_i[\overrightarrow{N}/\overrightarrow{x}] =_{\beta\eta} y_i$, a property of coordinates which cannot hold if s_i is bound. Similarly t_j is some y_i.

This provide us with two functions $\sigma : n \to m$, $\pi: m \to n$ such that

$$M_i = \lambda \overrightarrow{u}_i.x_{\sigma(i)} \overrightarrow{P}_i \text{ for } 1 \leq i \leq n, \qquad N_j = \lambda \overrightarrow{v}_j.y_{\pi(j)} \overrightarrow{Q}_j \text{ for } 1 \leq i \leq m.$$

In conclusion, for $1 \leq i \leq n$ we obtain[1]:

---

[1]We maintain here the original notation from [BDCL92] for delimiting the scope of a substitution: for example, $(P\{M[N/x]\})$ will stay for the term (PM) where the substitution $[N/x]$ is applied only to the subterm M.

$$y_i =_{\beta\eta} M_i[\vec{N}/\vec{x}]$$

$$=_{\beta\eta} \quad (\lambda\vec{u_i}.x_{\sigma(i)}\,\vec{P}_i)[\vec{N}/\vec{x}]$$

$$=_{\beta\eta} \quad \lambda\vec{u_i}.N_{\sigma(i)}\{\vec{P}_i[\vec{N}/\vec{x}]\}$$

$$=_{\beta\eta} \quad \lambda\vec{u_i}.(\lambda\vec{v}_{\sigma(i)}.y_{\pi(\sigma(i))}\,\vec{Q}_{\sigma(i)})\{\vec{P}_i[\vec{N}/\vec{x}]\}$$

$$=_{\beta\eta} \quad \text{if } \vec{v}_{\sigma(i)} \text{ is longer than } \vec{P}_i$$

$$\text{then } \lambda\vec{u_i}.\vec{v'}_{\sigma(i)}.y_{\pi(\sigma(i))}\,\vec{Q}_{\sigma(i)}[(\vec{P}_i[\vec{N}/\vec{x}])/(\overrightarrow{v_{\sigma(i)}} - \overrightarrow{v'_{\sigma(i)}})];$$

$$\text{otherwise } \lambda\vec{u_i}.y_{\pi(\sigma(i))}\{\vec{Q}_{\sigma(i)}[(\vec{P}_i[\vec{N}/\vec{x}])/\overrightarrow{v_{\sigma(i)}}]\}\{\vec{P'}_i[\vec{N}/\vec{x}]\}$$

In either case of the last equality, each term can reduce to y_i iff $y_i = y_{\pi(\sigma(i))}$ and each of the Q's and P's left reduces in the order to one of the bound variables, so that one can apply η, several times, at the end. The same holds for $N_j[\vec{M}/\vec{y}]$ for $1 \leq j \leq m$.

Thus $i = \pi(\sigma(i))$ for $1 \leq i \leq n$ and $j = \sigma(\pi(j))$ for $1 \leq j \leq m$, and we can conclude that $m = n$, σ is a permutation and π is its inverse. $\square$

Now, back to the second-order case, we can no longer assume that the coordinates are terms of the calculus $\lambda^2\beta\eta$. This means first of all that we must find a separate proof of the fact that the head-normal form of the M_i is $\lambda\vec{u_i}.x_{\sigma_i}\vec{P}_i$ (respectively, N_j is $\lambda\vec{v_j}.y_{p_j}\vec{Q}_j$), but this is not a serious problem. We can repair the proof and adapt it to the second-order case.

What really complicates matters is the need to prove now that the coordinates are terms of $\lambda^2\beta\eta$. Such a proof will need to examine inductively the full structure of coordinates, not only their head variable, so we will have to turn this lemma into a stronger one that provides us with an invariant to be used in this inductive proof. Let's see why.

Suppose we try to prove that the coordinates are terms of $\lambda^2\beta\eta$, by induction on their structure. The lemma tells us that any coordinate M_i is $\lambda\vec{u_i}.x_{\sigma_i}\vec{P}_i$, so we know that, say, the "prefix" $\lambda\vec{u_i}.x_{\sigma_i}$ of M_i can already be seen as a term of $\lambda^2\beta\eta$: seems nice! Unfortunately, here we immediately face a difficulty in applying the inductive hypothesis. We would like to say that by induction hypothesis the $\vec{P}_i$ are already terms of $\lambda^2\beta\eta$ and then conclude, but we could do this only in the case that the $\vec{P}$ and $\vec{Q}$ enjoy the properties of coordinates, while the lemmas stated above seems to tell us nothing at all about these terms. Apparently, we are stuck.

Looking closer at the proof of the lemma, though, we see that after all our original Lemma says something interesting, not about the P_i's or Q_j's, but about the $P_i[\vec{N}/\vec{x}]$'s and the $Q_i[\vec{M}/\vec{y}]$'s. These terms seem to enjoy the properties of coordinates, with respect to the variables v_i's and u_i's. In fact, if

$$Q_i[(\vec{P}_i[\vec{N}/\vec{x}])/\vec{u_i}] = u_i,$$

as shown in the proof of the lemma, then it is easy to see that also

$$Q_i[\vec{M}/\vec{y}][(\vec{P}_i[\vec{N}/\vec{x}])/\vec{u_i}] = u_i.$$

So the right candidates to be considered for an inductive argument will not be just the subterms P_i of the coordinates M_i but these subterms up to some substitution. We are then in the typical situation that needs an induction loading: to prove that the coordinates of an invertible term are terms of $\lambda^2\beta\eta$, we will prove something more, namely that terms that enjoy the properties of coordinates after some substitution is applied to them are all in $\lambda^2\beta\eta$. As a special case, we will have our theorem considering empty substitutions.

This brings up two more problems that need to be solved in order to make our new proof work:

- We must now modify the statement of the lemma to deal not just with coordinates but with coordinates to which is applied some substitution, in order for the inductive argument to apply to the case of $Q_i[\vec{M}/\vec{y}]$; furthermore, we must be able to deal with n-tuples containing also types and not only terms, as the Q_i's are not all terms, but can be types also.

- We must ensure that the number of the Q_i's and P_i's is the same, or, if it is not the case, find a way to extend the shorter one to the length of the longer, in a way as to turn them into coordinates enjoying the **distributed invertibility**.

The first problem is easily solved by modifying the statement of the lemma to handle not just coordinates (that enjoy the **distributed invertibility**) but coordinates to which is applied some kind of substitution. As for some of the Q_i's being types, our notion of **distributed invertibility** already takes care of them.

Definition 5.2.8 (head-free terms and substitutions)
A term M is called (second-order) head free when it has a head normal form

with free head variable, that is, its head normal form is $\lambda \vec{v}.x\vec{P}$ with x a free term variable (the abstractions can be both term and type abstractions). A head-free substitution is a substitution that replaces variables with head free terms and possibly type variables with types. We will use $\bullet$, $\circ$, $\vartriangleleft$, $\vartriangleright$, $\cdots$ to range over head free substitutions. Moreover, if $\bullet$ and $\circ$ are substitutions, $\bullet\circ$ will stand for the usual composition of substitutions, that is done right to left, that is, first apply $\bullet$ and then $\circ$.

Remark 5.2.9 *As suggested by the notation, the composition of two head-free substitutions is still a head-free substitution (see [Bar84], Lemma 21.2.3, p. 535).*

Actually, by looking at how the Q_i's relate to $\vec{M}$, we see that the M's are *terms*, with a free variable in head position, so the substitution we are interested in are actually head-free ones. Furthermore, the u_i's are distinct from the $\vec{x}$ and the M's do not contain free any variables from the N's or $\vec{y}$, so in our general definition we can require that the substitution behaves in the same way, that is, it affects only those free variables of $\vec{N}$ that are not also in $\vec{x}$. We can put all this together to get the definition of the distributed invertibility up to head-free substitutions.

Definition 5.2.10 (distributed invertibility up to substitutions)
Given $\vec{N}$, $\vec{M}$, $\vec{x}$, $\vec{y}$ as above, we will write $DistInv(\vec{M}, \vec{x}, \vec{N}, \vec{y})^{\bullet\circ}$ if $DistInv(\vec{M}^{\circ}, \vec{x}, \vec{N}^{\bullet}, \vec{y})$ holds, where

1. *the head-free substitution $\circ$ affects only variables that are in $(FV(\vec{N}) \setminus \vec{x})$ and maps them into terms with no occurrences of $FV(\vec{N})$ or $\vec{y}$ or type expressions*

2. *the head-free substitution $\bullet$ affects only variables that are in $(FV(\vec{M}) \setminus \vec{y})$ and maps them into terms with no occurrences of $FV(\vec{M})$ or $\vec{x}$ or type expressions*

3. *these head-free substitutions $\circ$ and $\bullet$ substitute type expressions with no occurrences of $\times$ or $\mathbf{T}$ for type variables.*

Using this definition of distributed invertibility in the modified lemma, we will be able to overcome the first problem described above.

As for the second problem, namely that the number of the Q_i's and P_i's must be the same in order to apply induction, we can actually extend as needed the shorter sequence to match the longer one by means of fresh variables (this will be shown in the proof of the modified lemma).

Now, we are almost done. Let's put it all together, and see how the modified lemma looks.

Lemma 5.2.11 (Main Lemma)

Let $\overrightarrow{N}$, $\overrightarrow{M}$ (with $m = |\overrightarrow{N}|$ and $n = |\overrightarrow{M}|$) be sequences of terms and/or types and $\circ$, $\bullet$ be head-free substitutions such that $DistInv(\overrightarrow{M}, \overrightarrow{x}, \overrightarrow{N}, \overrightarrow{y})^{\overset{\circ}{\bullet}}$, then the following hold:

1. *the substitutions $\bullet$ and $\circ$ are idempotent, i.e., $\bullet\bullet = \bullet$ and $\circ\circ = \circ$*

2. *when M_i and N_j are terms,*

$$M_i = \lambda\overrightarrow{v_i}.x_{\sigma(i)}\overrightarrow{P_i}, \quad N_j = \lambda\overrightarrow{u_j}.y_{\pi(j)}\overrightarrow{Q_j}$$

   where $\sigma{:}n \to m$, $\pi{:}m \to n$ are integer functions s.t. $\sigma(\pi(i))=i$ if M_i is a term and $\pi(\sigma(j))=j$ if N_j is a term

3. *$n = m$*

4. *every P_{i_k} (Q_{j_h}) that is a type expression is just a type variable*

5. *every variable free in P_{i_k} (Q_{j_h}) has no occurrence of $\times$ or $\mathbf{T}$ in the type expressions occurring in it*

6. *P_{i_k} (Q_{j_h}) has no occurrence of $\times$ or $\mathbf{T}$ in its type if P_{i_k} (Q_{j_h}) is a term*

7. *Define $s_1 = |\overrightarrow{P_i}|, r_2 = |\overrightarrow{u}_{\sigma(i)}|, r_1 = |\overrightarrow{v}_i|, s_2 = |\overrightarrow{Q}_{\sigma(i)}|$. Without loss of generality, suppose $s_1 > r_2$. Then $r_1 > s_2$. Furthermore, $\overrightarrow{Q}_{\sigma(i)}$ can be extended with a sequence of type and/or term variables $[u''_1, \cdots, u''_{r_1-s_2}]$ to a sequence $\overrightarrow{Q'}_{\sigma(i)}$ such that*

$$DistInv(\overrightarrow{P}_i, \overrightarrow{v}_i, \overrightarrow{Q'}_{\sigma(i)}, \overrightarrow{u'}_{\sigma(i)})^{\overset{\triangleright\triangleleft}{}} \quad \text{holds, where } \overrightarrow{u'}_{\sigma(i)} = \overrightarrow{u}_{\sigma(i)} \cup [u''_1, \cdots, u''_{r_1-s_2}] \text{ and } \triangleleft, \triangleright \text{ are suitable head-free substitutions.}$$

Proof. This is done by a tedious case analysis (see Appendix B for full details). $\square$

Remark 5.2.12 *The original lemma had just conditions 2 and 3. Conditions 5, 6 and 7 are needed to allow the argument to inductively apply to the subterms, as described before, while the remaining conditions 1 and 4 are needed for technical reasons to make the final proof work.*

The reader familiar with the original characterization from [Dez76] will notice how the proof of this lemma has a very similar flavor, but with the further complications arising from typing, which does not allow arbitrary η expansions and forces us to introduce condition 7.

Relating coordinates to invertible terms in $\lambda^2 \beta \eta$

Now, condition 7 is exactly what is needed for our induction argument. We start with a set of coordinates up to substitution, and we end with a set of smaller coordinates, still up to substitution, so that the following Proposition can be very easily proved by induction.

Proposition 5.2.13 *If $DistInv(\overrightarrow{M}, \overrightarrow{x}, \overrightarrow{N}, \overrightarrow{y})^{\bullet\circ}$, then*

1. *the terms in $\overrightarrow{N}$ and $\overrightarrow{M}$ are terms of $\lambda^2 \beta \eta$, and*

2. *every occurrence of a type expression is just a type variable.*

Proof. By an easy induction on the complexity n of the longest term in $\overrightarrow{N}$ and $\overrightarrow{M}$.

- Base:

  (1) $n = 1$ so every term in $\overrightarrow{N}$ and $\overrightarrow{M}$ is just a variable (cannot be $* : \mathbf{T}$: remember that the type of the elements of the sequences are simple, hence do not contain $\mathbf{T}$)

  (2) because of point 3 in the definition of $DistInv(\overrightarrow{M}, \overrightarrow{x}, \overrightarrow{N}, \overrightarrow{y})^{\bullet\circ}$, (i.e., $M_i[\overrightarrow{N}/\overrightarrow{x}] = y_i$ and $N_i[\overrightarrow{M}/\overrightarrow{y}] = x_j$), every type expression has to be a simple type variable. There are no other type expressions as the terms are of length 1.

- Ind. Step: $n + 1$

  (1) We get from the main lemma that $M_i = \lambda \overrightarrow{v_i}.x_s(i)\overrightarrow{P_i}$ and $N_j = \lambda \overrightarrow{u_j}.y_p(j)\overrightarrow{Q_j}$ with $DistInv(\overrightarrow{P}i, \overrightarrow{v_i}, Q'_{\sigma(i)}, \overrightarrow{u}'_{\sigma(i)})^{\bowtie}$, and where the complexity of the terms in $\overrightarrow{P_i}$ and $\overrightarrow{Q'}_{\sigma(i)}$ is strictly lower than $n + 1$ for every i and j. We can apply the induction hypothesis and get that every term in $\overrightarrow{P_i}$ and $\overrightarrow{Q'}_{\sigma(i)}$ belongs to $\lambda^2 \beta \eta$ too. Furthermore, we know from the main lemma that every type expression in $\overrightarrow{P_i}$ and $\overrightarrow{Q'}_{\sigma(i)}$ is just a type variable. Since $\overrightarrow{Q'_{\sigma(i)}}$ $\subseteq \overrightarrow{Q'_{\sigma(i)}}$, this suffices to show that every $M_i = \lambda \overrightarrow{v_i}.x_s(i)\overrightarrow{P_i}$ and $N_j = \lambda \overrightarrow{u_j}.y_{\pi(j)}\overrightarrow{Q_j}$ has no occurrence of constants or complex type expression in it and belongs to $\lambda^2 \beta \eta$.

(2) as in the base case for elements in $\overrightarrow{N}$ and $\overrightarrow{M}$ that are type expressions. Direct consequence of 1 for type expressions occurring inside terms.

$\square$

And we get, as a special case,

Corollary 5.2.14 *The coordinates* $\overrightarrow{N} = [N_1, \cdots, N_m]$ *and* $\overrightarrow{M} = [M_1, \cdots, M_n]$ *in Proposition 5.2.4 are terms of* $\lambda^2\beta\eta$. *Furthermore,* $n = m$.

Proof. It follows from the definitions that $DistInv(\overrightarrow{M}, \overrightarrow{x}, \overrightarrow{N}, \overrightarrow{y})^{\bullet\circ}$ holds with $\bullet$ and $\circ$ empty substitutions. Hence, the result follows from the previous proposition, point 1, as well as $n = m$. $\square$

So we have factored out pairing and the constant $*$, and we have restricted ourselves to $\lambda^2\beta\eta$. We can do the same for equalities.

Lemma 5.2.15 *For the coordinates as in Proposition 5.2.4, the equalities* $M_i[\overrightarrow{N}/\overrightarrow{x}]=_{\beta^2\eta^2\pi*} y_i$ *and* $N_i[\overrightarrow{M}/\overrightarrow{y}]=_{\beta^2\eta^2\pi*} x_j$ *hold in* $\lambda^2\beta\eta$ *(that is, we have that* $M_i[\overrightarrow{N}/\overrightarrow{x}]=_{\beta^2\eta^2} y_i$ *and* $N_i[\overrightarrow{M}/\overrightarrow{y}]=_{\beta^2\eta^2} x_j$ *).*

Proof. Since the reduction $\overset{\beta^2\eta^2\pi*}{\longrightarrow}$ is Church-Rosser and a variable is in normal form (notice that the variables we are considering are not redexes for *gentop* as their types do not contain $\mathbf{T}$), there must be a reduction path from $M_i[\overrightarrow{N}/\overrightarrow{x}]$ to y_i. We know from the previous Corollary 5.2.14 that there is no constant in the terms we consider, so there is no π, SP, *gentop*, η_{top} or SP_{top} redex nor any can be created by any reduction. So this reduction path must be made of β and η reductions only, which implies that equality holds for the system consisting of β plus η alone too. This is a special case of a more general phenomenon, as seen in Proposition 2.6.2. $\square$

5.2.4 Syntactic characterization of canonical bijections

We have shown that the coordinates of a canonical bijection are terms of $\lambda^2\beta\eta$, and we also know now that these coordinates, even when seen as terms of $\lambda^2\beta\eta$, still enjoy this peculiar property we called **distributed invertibility**. We are now in a position to relate these coordinates to invertible terms of the pure $\lambda^2\beta\eta$. As we have already seen in the survey, a characterization of the invertible terms of $\lambda^2\beta\eta$ is provided in [BL85] as the

2-f.h.p. (see 1.9.6 and 1.9.7). Now, it is easy to check that the coordinates M_i of a canonical bijection are the *body* of a 2-f.h.p. (in the sense that they contain only a free variable $x_{\sigma(i)}$, such that $\lambda x_{\sigma(i)}.M_i$ is a 2-f.h.p.).

Theorem 5.2.16 *For the coordinates in Proposition 5.2.4 there exist a permutation $\sigma{:}n \to n$ s.t. $\lambda x_{\sigma(i)}.M_i$ and $\lambda y_{\pi(j)}.N_j$ are second-order f.h.p.'s, where π is σ^{-1}.*

Proof. Just build the $\lambda^2\beta\eta$ terms, for a suitably typed new variable z,

$$M = \lambda z x_1 \cdots x_n.z M_1 \cdots M_n, \quad N = \lambda z y_1 \cdots y_n.z N_1 \cdots N_n$$

They are terms of $\lambda^2\beta\eta$ (Corollary 5.2.14), and it is easy to check that they are invertible w.r.t. $\beta\eta$ equality, as

$$
\begin{aligned}
M \circ N \quad &=_{\beta^2\eta^2} \quad \lambda w.(M(Nw)) \\
&=_{\beta^2\eta^2} \quad \lambda w.(\lambda z x_1 \cdots x_n.z M_1 \cdots M_n)((\lambda z y_1 \cdots y_n.z N_1 \cdots N_n)w) \\
&=_{\beta^2\eta^2} \quad \lambda w.(\lambda z x_1 \cdots x_n.z M_1 \cdots M_n)(\lambda y_1 \cdots y_n.w N_1 \cdots N_n) \\
&=_{\beta^2\eta^2} \quad \lambda w.(\lambda x_1 \cdots x_n.(\lambda y_1 \cdots y_n.w N_1 \cdots N_n) M_1 \cdots M_n) \\
&=_{\beta^2\eta^2} \quad \lambda w.\lambda x_1 \cdots x_n.w N_1 \cdots N_n[\overrightarrow{M_i}/\overrightarrow{y_i}] \\
&=_{\beta^2\eta^2} \quad \lambda w.\lambda x_1 \cdots x_n.w N_1[\overrightarrow{M_i}/\overrightarrow{y_i}] \cdots N_n[\overrightarrow{M_i}/\overrightarrow{y_i}] \\
&=_{\beta^2\eta^2} \quad \lambda w.\lambda x_1 \cdots x_n.w x_1 \cdots x_n \text{ (due to Lemma 5.2.15)} \\
&=_{\beta^2\eta^2} \quad \lambda w.w.
\end{aligned}
$$

Similarly for $N \circ M$.

So N and M are second-order f.h.p. and this implies that every M_i has only one occurrence of the x_i's (namely $x_{\sigma(i)}$) and the same for the N_i's. Hence

$$
\begin{aligned}
M_i[\overrightarrow{N_i}/\overrightarrow{x_i}] \quad &\equiv \quad M_i[N_{\sigma(i)}/x_{\sigma(i)}] =_{\beta^2\eta^2} y_i, 1 \le i \le n \\
N_i[\overrightarrow{M_i}/\overrightarrow{y_i}] \quad &\equiv \quad N_i[M_{\pi(i)}/y_{\pi(i)}] =_{\beta^2\eta^2} x_i, 1 \le i \le n
\end{aligned}
$$

and

$$\lambda x_{\sigma(i)}.M_i : A_{\sigma(i)} \to B_i, \quad \lambda y_i.N_{\sigma(i)} : B_i \to A_{\sigma(i)}$$

are second-order f.h.p. $\square$

Now, this exact knowledge of the shape of a coordinate finally gives us the syntactic characterization of canonical bijections.

Theorem 5.2.17 (shape of a canonical bijection)
Let $M : A_1 \times \cdots \times A_n \to B_1 \times \cdots \times B_n$ be a canonical bijection of $\lambda^2 \beta \eta \pi *$. Then either M is the identity or there is a permutation $\pi . n \to n$ s.t.

1. $n.f.(M) = \lambda z.(\langle M_1[(\mathrm{p}_{\pi(1)}z)/x_{\pi(1)}], \cdots, M_n[(\mathrm{p}_{\pi(n)}z)/x_{\pi(n)}]\rangle)$,

2. $\lambda x_{\pi(1)}.M_i : A_{\pi(i)} \to B_i$ are 2-f.h.p.'s

3. in particular, $M = \lambda z.(\langle M_1'(\mathrm{p}_{\pi(1)}z), \cdots, M_n'(\mathrm{p}_{\pi(n)}z)\rangle)$, with $M_i' = \lambda x_{\pi(i)}.M_i$'s 2-f.h.p.'s

Proof. We have shown in Proposition 5.2.4 that every canonical invertible term M can be written in terms of its coordinates M_i's as

$$\lambda z.(\lambda x_1 \cdots x_n.(\langle M_1, \cdots, M_n \rangle))(\mathrm{p}_1 z) \cdots (\mathrm{p}_n z).$$

This term reduces to $\lambda z.(\langle M_1[(\mathrm{p}_{\pi(1)}z)/x_{\pi(1)}], \cdots, M_n[(\mathrm{p}_{\pi(n)}z)/x_{\pi(n)}]\rangle)$, which we claim is in normal form if it does not reduce via SP to $\lambda z.z$. This follows immediately from the fact that the coordinates are (by construction) in normal form and that the substitutions $[(\mathrm{p}_{\pi(i)}z)/x_{\pi(i)}]$ do not create any new redex. So we get 1.

Theorem 5.2.16 guarantees 2.

Setting now $M_i' = \lambda x_{\pi(1)}.M_i$, we get 3.

Notice that in the particular case $n = 1$, $nf(M) = \lambda z.M_1$ and it is a 2-f.h.p. $\square$

Corollary 5.2.18 *It is possible to decide if a generic term $M : A \to B$ is invertible.*

Proof. Build the term $M' : n.f(A) \to n.f.(B)$ as in Proposition 5.1.6. Then normalize it and check if its shape is as in Theorem 5.2.17. $\square$

Now we can use this characterization to prove completeness of $Th_{\times T}^2$ for isomorphisms of type normal forms given by canonical bijections (and hence for all isomorphisms of types, Propositions 5.1.7 and 5.1.8) in the following section.

5.3 Completeness for isomorphisms

We know now that two types A and B are isomorphic iff their normal forms are; furthermore, $Th_{\times T}^2$ equates a type to its normal form. Thus it is possible to show that $Th^2_{\times \mathbf{T}}$ is complete for the definable isomorphisms if we can show that $Th^2_{\times \mathbf{T}}$ is complete for isomorphisms between regular types. But if two regular types are definably isomorphic, then the isomorphism can be defined by a canonical invertible term, and we can easily get the result by inspecting its structure, as given in

Proposition 5.3.1 *Let A, B be regular type expressions that are not $\mathbf{T}$. Then $A \cong_d B \Rightarrow Th^2_{\times \mathbf{T}} \vdash A = B$.*

Proof. Suppose $A \cong_d B$ via a canonical invertible term M. We can assume that free and bound type variables are all different in M, as the theory $Th^2_{\times \mathbf{T}}$ allows renaming of bound type variables. If $A = A_1 \times \cdots \times A_m$, then $B = B_1 \times \cdots \times B_m$ (as the length of two isomorphic regular types is the same, see Corollary 5.2.14).

If $m = 1$ then $A \cong_d B$ via a 2-f.h.p. M and we will show the completeness of $Th^2_{\times \mathbf{T}}$ by induction on the depth of the Böhm-tree $BT(M)$ of M.

- Depth 1: $M \equiv \lambda z : C.z$. Thus $M : C \to C$, and $Th^2_{\times \mathbf{T}} \vdash C = C$ by reflexivity.

- Depth $d + 1$: $M \equiv \lambda z : A.\lambda \overrightarrow{x} : \overrightarrow{B}.z\overrightarrow{N}$. Recall $z\overrightarrow{N} = zN_1 \cdots N_n$ where, for some permutation σ, if the ith abstraction in $\lambda x : D$ is $\lambda x_i : D_i$ then $\lambda x_{\sigma(i)} : D_i.N_i$ is a 2-f.h.p. We proceed by induction on the length n of $\overrightarrow{N}$. If $n = 0$ then the result follows as in the case of depth 1. If $n = k + 1$, then $M : A \to B$, with

$$M \equiv \lambda z : A.\lambda x_1 : B_1 \cdots \lambda x_{\sigma(k+1)} : B_{\sigma(k+1)} \cdots \lambda x_{k+1} : B_{k+1}.zN_1 \cdots N_{k+1}.$$

In what follows, $A \,\square\, B$ will stand for either $A \to B$ or $\forall A.B$. First, we notice that, in order to type-check, we must have

$$A = (A_1 \square_1 \cdots \square_k A_{k+1} \square_{k+1} E) \text{ for some } E,$$

$$B = (B_1 \square_1 \cdots \square_k B_{k+1} \square_{k+1} F) \text{ for some } F,$$

with $F = E[N_{i_1}/A_{i_1}] \cdots [N_{i_r}/A_{i_r}]$, where the i_j are all the indexes s.t. $\square_{i_r} = \forall$ in A.

Then, we proceed by cases on N_{k+1}:

1. N_{k+1} is a term. Then we have

    (i) $\lambda x_{\sigma(k+1)} : B_{\sigma(k+1)}.N_{k+1} : B_{\sigma(k+1)} \to Type(N_{k+1})$

    (ii) $\square_{\sigma(k+1)} = \to$

As $\mid BT(\lambda x_{\sigma(k+1)} : B_{\sigma(k+1)}.N_{k+1}) \mid < d + 1$, we know by induction hypothesis on d that

$$(iii) \qquad Th^2_{\times \mathbf{T}} \vdash B_{\sigma(k+1)} = Type(N_{k+1}).$$

Furthermore, it is straightforward to see that, since free and bound type variables in M are different, the term

$$\lambda z : A.\lambda x_1 : B_1 \cdots \lambda x_{\sigma(k+1)-1} : B_{\sigma(k+1)-1}.$$
$$\lambda x_{\sigma(k+1)+1} : B_{\sigma(k+1)+1} \cdots \lambda x_{k+1} : B_{k+1}.zN_1 \cdots N_k : A \to B'$$

is well formed and is still a 2-f.h.p., with only k N's. Then, by induction hypothesis on n,

$$(iv) \qquad Th^2_{\times \mathbf{T}} \vdash A = B'.$$

Now notice that

$$B' = (B_1 \square_1 \cdots B_{\sigma(k+1)-1} \square_{\sigma(k+1)-1} B_{\sigma(k+1)+1} \square_{\sigma(k+1)+1}$$
$$\cdots \square_k B_{k+1} \square_{k+1} Type(N_{k+1}) \to F).$$

So, since equality is substitutive, (ii), (iii) and (iv) yield

$$Th^2_{\times \mathbf{T}} \vdash A = B'$$
$$= (B_1 \square_1 \cdots B_{\sigma(k+1)-1} \square_{\sigma(k+1)-1} B_{\sigma(k+1)+1} \square_{\sigma(k+1)+1}$$
$$\cdots \square_k B_{k+1} \square_{k+1} B_{\sigma(k+1)} \to F)$$
$$= (B_1 \square_1 \cdots B_{\sigma(k+1)-1} \square_{\sigma(k+1)-1} B_{\sigma(k+1)+1} \square_{\sigma(k+1)+1}$$
$$\cdots \square_k B_{k+1} \square_{k+1} B_{\sigma(k+1)} \square_{\sigma(k+1)} F).$$

Now, it suffices to show that this last type is equal to B in $Th^2_{\times \mathbf{T}}$. Since we assumed free and bound type variables to be different, if $\square_j$ is a $\forall$, with j greater than $\sigma(k+1)$, then B_j cannot be free in $B_{\sigma(k+1)}$. This allows us to use axiom 10, together with 8 and the equality $A \to (B \to C) = B \to (A \to C)$ derived from axioms 1 and 3 (that are always applicable), in order to repeatedly swap $B_{\sigma(k+1)} \square_{\sigma(k+1)}$ with $B_j \square_j$ up to $B_{\sigma(k+1)+1} \square_{\sigma(k+1)+1}$. By this, we obtain the required equality.

2. N_{k+1} is a type. Then we have $M : A \to B$, with

$$M \equiv \lambda z : A.\lambda x_1 : B_1 \cdots \lambda X_{\sigma(k+1)} \cdots \lambda x_{k+1} : B_{k+1}.zN_1 \cdots N_k[X_{\sigma(k+1)}].$$

As in the previous case, we can reduce to a smaller k via the term

$$\lambda z : A.\lambda x_1 : B_1 \cdots \lambda x_{\sigma(k+1)-1} : B_{\sigma(k+1)-1}.$$
$$\lambda x_{\sigma(k+1)+1} : B_{\sigma(k+1)+1} \cdots \lambda x_{k+1} : B_{k+1}.zN_1 \cdots N_k : A \to B'$$

which is well formed, as free and bound variables are different, so that removing the abstraction does not cause any capture of variables by other type abstractions. It is still a 2-f.h.p., so we get $Th^2_{\times\mathbf{T}} \vdash A = B'$. Now,

$$B' = (B_1 \square_1 \cdots B_{\sigma(k+1)-1} \square_{\sigma(k+1)-1} B_{\sigma(k+1)+1} \square_{\sigma(k+1)+1}$$
$$\cdots \square_k B_{k+1} \square_{k+1} \forall Z.F')$$

where $F \equiv F'[X_{\sigma(k+1)}/Z]$, so that $\forall X_{\sigma(k+1)}.F \equiv \forall X_{\sigma(k+1)}.F'[X_{\sigma(k+1)}/Z]$, which is equal, by axiom 9, to $\forall Z.F'$. Hence,

$$Th^2_{\times\mathbf{T}} \vdash A = B'$$
$$= (B_1 \square_1 \cdots B_{\sigma(k+1)-1} \square_{\sigma(k+1)-1} B_{\sigma(k+1)+1} \square_{\sigma(k+1)+1}$$
$$\cdots \square_k B_{k+1} \square_{k+1} \forall Z.F')$$
$$= (B_1 \square_1 \cdots B_{\sigma(k+1)-1} \square_{\sigma(k+1)-1} B_{\sigma(k+1)+1} \square_{\sigma(k+1)+1}$$
$$\cdots \square_k B_{k+1} \square_{k+1} \forall X_{\sigma(k+1)}.F)$$
$$= (B_1 \square_1 \cdots B_{\sigma(k+1)-1} \square_{\sigma(k+1)-1} B_{\sigma(k+1)+1} \square_{\sigma(k+1)+1}$$
$$\cdots \square_k B_{k+1} \square_{k+1} B_{\sigma(k+1)} \square_{\sigma(k+1)} F).$$

To show that this last type is equal to B in $Th^2_{\times\mathbf{T}}$, we can use the swap axioms 8 and 10, but the latter is a conditional axiom, which can be used as we want only if we can ensure that $B_{\sigma(k+1)} \equiv X_{\sigma(k+1)}$ is not free in any B_j s.t. $\square_j$ is $\to$.

Indeed, we can show this fact, proceeding by contradiction. Suppose that $X_{\sigma(k+1)}$ is free in some B_j s.t. $\square_j$ is $\to$. Notice that the types equated by the theory $Th^2_{\times\mathbf{T}}$ have the same free type variables, so that $N_{\sigma^{-1}(j)}$ is then a term, and $X_{\sigma(k+1)}$ is free in its type, which is isomorphic to B_j as already remarked. For M to type-check, then, the type of $zN_1 \cdots N_{(\sigma^{-1}(j))-1}$ must be $Type(N_{\sigma^{-1}(j)}) \to G$ for some G.

Now notice that, due to the structure of 2-f.h.p.'s, $X_{\sigma(k+1)}$ occurs (in type applications) only once and exactly as N_{k+1} in M: under these conditions, we can prove by induction on k that $X_{\sigma(k+1)}$ must occur in the type of z. This leads to a contradiction, though, as then the subterm

$$\lambda X_{\sigma(k+1)} \cdots \lambda x_{k+1} : B_{k+1}.zN_1 \cdots N_k[X_{\sigma(k+1)}]$$

of M is not well formed. There is a type abstraction over a type variable $(X_{\sigma(k+1)})$ occurring free in the type of a free term variable (z), and this violates the term formation rule for type abstraction of Definition 1.5.1.

If $m > 1$, then by Theorem 5.2.17 there exist an integer m and a permutation $\pi : m \to m$ s.t. the normal form of M is

$$\lambda z.(\langle M_1[(\mathrm{p}_{\pi(1)}z)/x_p(1)], \cdots, M_n[(\mathrm{p}_{\pi(m)}z)/x_p(m)]\rangle),$$

where the terms $\lambda x_i.M\pi(i) : A_i \to B_{\pi(i)}$ are second-order f.h.p.'s. This means that $A_i \cong_d B_{\pi(i)}$ via 2-f.h.p.'s, which are invertible. We have already shown that $Th^2_{\times \mathbf{T}}$ is complete for isomorphisms definable by 2-f.h.p.'s, and to get the result it is enough to notice that $Th^2_{\times \mathbf{T}}$ includes commutativity and associativity for the product. $\square$

So we can finally get our main theorem.

Theorem 5.3.2 (main theorem: difficult implication)
$A \cong B \Rightarrow Th^2_{\times \mathbf{T}} \vdash A = B.$

Proof. By Theorem 1.8.3, it is enough to show that $A \cong_d B \Rightarrow Th^2_{\times \mathbf{T}} \vdash A = B$. This is now an easy consequence of Propositions 5.1.7 and 5.3.1. $\square$

5.3.1 Uniform isomorphisms

This allows us to conclude that an isomorphism valid in every model can be expressed by means of an uniform bijection, which is not at all obvious in general.

Corollary 5.3.3 *If $A \cong B$, then there is an uniform isomorphism between A and B.*

Proof. If A and B are isomorphic in every model, then they are provably equal in $Th^2_{\times \mathbf{T}}$; since for every axiom of $Th^2_{\times \mathbf{T}}$ there is an uniform isomorphism, we can derive a uniform isomorphism for B and A by just composing the ones associated to every step of the proof of $B = A$. $\square$

△ 5.4 Decidability of the equational theory

An immediate consequence of the main theorem is

Theorem 5.4.1 *Given types A and B it is decidable whether they are isomorphic in all models of $\lambda^2 \beta \eta \pi *$.*

Proof. Let the type normal-forms of A and B be $(A_1 \times \cdots \times A_n)$ and $(B_1 \times \cdots \times B_m)$, respectively. We know that none of the A_i and B_j can contain any occurrence of $\mathbf{T}$ or $\times$ (by Remark 5.1.4). Now we know that A and B are isomorphic if and only if $n = m$ (Lemma 5.2.14), and there exist a permutation $\sigma : n \to n$ such that $A_i \cong_d B_{\sigma(j)}$. Hence, to decide $A \cong_d B$ it suffices to be able to decide $A_i \cong_d B_j$, as then we can just try $A_i \cong_d B_{\sigma(i)}$ for all $1 \leq i \leq n$ and all permutations $\sigma : n \to n$.

By inspecting the proof of Proposition 5.3.1, we see that to axiomatize $A_i \cong_d B_{\sigma(i)}$, we only need the equational theory (that we will call S) made up of axioms 8, 9, 10 and the equality $A \to (B \to C) = B \to (A \to C)$ derived from axioms 1 and 3.

These equalities do not change the length of formulae (if we use the notation of Proposition 5.3.1); hence, if $S \vdash A_i = B_j$, then A_i and B_j have the same length.

Axiom 9, though, changes the alphabet of a formula, so that we need to gain some control on its application in the proofs of equality modulo S if we want to provide an effective decision procedure. Actually, if we start with two types A and B equal in S where free and bound type variables are different (and we can do it without loss of generality), we can find a proof of equality where any use of axiom 9 appears only at the end as follows:

- First, given a proof $A = A_1 = \cdots = A_n = B$, we transform it into a proof $A = A'_1 = \cdots = A'_n = B$ where at every stage all free and bound variables are distinct.[2] It suffices, whenever a variable X is replaced by Y via axiom 9 in the proof, to replace Y by a fresh variable Z never used in the proof. It is easy to see that the resulting sequence of formulae is still a proof in S, since the only constraint on the name of the variables appears when using axiom 10 from right to left. And if a bound occurrence of Y is not free in some formula C outside the scope of the $\forall$ that binds Y, then *a fortiori* Z is not free in C, by the way it has been chosen. The proof obtained after applying orderly this proceed to every occurrence of axiom 9 is the required proof $A = A'_1 = \cdots = A'_n = B$.

- Second, in a proof where all the free and bound variables are distinct, we can push all the applications of axiom 9 at the end. This can be shown by noticing that a sequence of equalities $A_{i-1} =_{axiom\ 9} A_i = A_{i+1}$, where the second equality is not an application of axiom 9 can be always replaced by a sequence $A_{i-1} = A'_i =_{axiom\ 9} A_{i+1}$,

---

[2] Actually, as remarked by one of the referees, what we obtain is not exactly B but an α variant of B, as we may rename bound variables in our procedure. In that case, we can easily complete the proof with some steps of α equality, so we need not consider this case explicitly in the proof.

where the first equality is not an application of axiom 9 and free and bound variables are all distinct in A'_i too. The only nontrivial case is axiom 10 after axiom 9, but under the hypothesis of having all distinct variables, everything works fine.

This means that if $A =_S B$, then B is just an α variant of one of the formulae equal to A in S less axiom 9. These formulae are in finite number (as these last axioms do not change the length nor the alphabet of a formula) and can be effectively generated (for example, by a depth-first search of the equality proofs). So it suffices to check B against each of these for α equality, which is a decidable task. $\square$

Since $Th^2_{\times T}$ is the theory of isomorphic types, an immediate consequence of the theorem above is the decidability of $Th^2_{\times T}$.

Corollary 5.4.2 *Equality in $Th^2_{\times T}$ is decidable.*

Deciding equality in theories containing associative and commutative operators, as well as binding operators, like $\forall$, is in general far from trivial (actually it is at least as hard as deciding Graph Isomorphism, see [Bas90]), so that the complexity of the decision procedure given above (it essentially requires to examine all the search space) is no surprise. Actually, the very fact that $Th^2_{\times T}$ is decidable is far from being an obvious result, and the insight gained by the study of invertible terms (in Proposition 5.3.1) is essential in order to establish it.

5.5   The complete theories of $\lambda^2\beta\eta\pi$ and $\lambda^2\beta\eta*$

Here we found, even if a little hidden by the more complex details of the proofs, the very same pattern as for the first-order system: indeed, the work done for $Th^2_{\times T}$ yields us immediately the complete theories of isomorphisms for $\lambda^2\beta\eta*$ and $\lambda^2\beta\eta\pi$.

Definition 5.5.1 $Th^2_\times$, the complete theory of isomorphic types of $\lambda^2\beta\eta\pi$, is a theory of equality plus the following axiom schemas:

1. $A \times B = B \times A$

2. $A \times (B \times C) = (A \times B) \times C$

3. $(A \times B) \to C = A \to (B \to C)$

4. $A \to (B \times C) = (A \to B) \times (A \to C)$

5. $\forall X.\forall Y.A = \forall Y.\forall X.A$

6. $\forall X.A = \forall Y.A[Y/X]$ (X free for Y in A, $Y \notin FTV(A)$)

7. $\forall X.(A \to B) = A \to \forall X.B$ ($X \notin FTV(A)$)

8. $\forall X.A \times B = \forall X.A \times \forall X.B$

Indeed, all these axioms are valid for $\lambda^2\beta\eta\pi$, and during the proof of completeness, no other axioms are used if the original types do not involve **T**, so there is no further isomorphism in $\lambda^2\beta\eta\pi$ (or otherwise there would be also in $\lambda^2\beta\eta\pi*$, which is not the case).

On the other side, if it is the product type-constructor that is omitted, then we cannot use any axiom related to the product. Anyway, axiom **Swap** is no longer derivable in the absence of axioms involving products, so one needs to add it explicitly, and we get the complete theory of isomorphisms for $\lambda^2\beta\eta*$ as follows:

Definition 5.5.2 $Th_{\mathbf{T}}^2$, the complete theory of isomorphic types of $\lambda^2\beta\eta*$, is a theory of equality plus the following axiom schemas:

1. $A \to (B \to C) = B \to (A \to C)$ (**Swap**)

2. $A \to \mathbf{T} = \mathbf{T}$

3. $\mathbf{T} \to A = A$

4. $\forall X.\mathbf{T} = \mathbf{T}$

5. $\forall X.\forall Y.A = \forall Y.\forall X.A$

6. $\forall X.A = \forall Y.A[Y/X]$ (X free for Y in A, $Y \notin FTV(A)$)

7. $\forall X.(A \to B) = A \to \forall X.B$ ($X \notin FTV(A)$)

Again, all these axioms are valid for $\lambda^2\beta\eta*$, and during the proof of completeness no other axioms are used if the original types do not involve products, so there is no further isomorphism in $\lambda^2\beta\eta*$ (or otherwise there would be also in $\lambda^2\beta\eta\pi*$, which we know it is not the case).

5.6   Conclusions

We have provided in the last two chapters a finite, complete and decidable axiomatization of the types isomorphic in every model of all the interesting subsystems (first- and second-order) of $\lambda^2\beta\eta\pi*$, by means of purely syntactical proof theoretic methods. The first-order case could be handled rather easily due to the pretty reduced interaction between types and terms that characterizes the monomorphic systems, while for the second-order case we had to deploy the full power of our techniques in order to overcome the difficulties posed by polymorphic types. It would also be interesting to investigate how far our methods can be extended to yield similar results for other or further extensions of the λ-calculus, to provide similar characterizations in the presence of recursive or inductive types.

Due to the connection between explicitly typed lambda calculus and intuitionistic logic, this work also fully characterizes the *constructively equivalent* formulae of IPC($\forall$, $\Rightarrow$, $\wedge$, True), the second-order intuitionistic positive propositional calculus, as is described in the introduction.

On the practical side, these results give the necessary theoretical basis to the development both of library search tools based on the *type as specification* paradigm and of extensions of the usual type-checking algorithms for strongly typed functional languages, that we will tackle in the next chapter. As we will see, we are in for some surprises, but our tools are refined enough by now to handle also the case of implicitly typed systems, like ML, where our work will finally find its immediate practical application.

Appendixes

Here follow three appendixes that contain in full detail most of the technicalities that are not immediately necessary to follow the proof of completeness. The first one provides some interesting result about n-tuples, the second one contains the very technical proof of the main lemma, and the last one mostly contains direct proofs of statements that are established in an indirect way in the main body of the chapter.

A   Properties of n-tuples

Notation A.1 (n-fold projections) *Let* $M : A_1 \times \cdots \times A_n$ *where the* $A_i's$ *have no occurrence of product or* **T**. *We will then write* $\mathrm{p}_i^k M$ *to note, if* $k \leq n, 0 < i < k$, *the term*

$$\underbrace{\mathrm{p}_1 \mathrm{p}_2 \cdots \mathrm{p}_2}_{i-1} M,$$

and, if $k \leq n, 1 < i = k$, *the term*

$$\underbrace{\mathrm{p}_2 \cdots \mathrm{p}_2}_{i-1} M.$$

The idea behind this notation is that a sequence of binary projections can be considered as a projection over an n-tuple. Then p_i^k is a notation for the sequence of projections that selects the ith component in a k-tuple. Obviously, any n-tuple is also a k-tuple if $k \leq n$, so one could be tempted to drop the suffix k from p_i^k, but this is not correct: the kth component of an n-tuple $\langle M_1, \cdots, M_n \rangle$ is M_k, which is not at all the same thing as the last component of the same term when considered as a k-tuple, which is $\langle M_k, \cdots, M_n \rangle$. The sequence of projections needed to select the ith component of a term considered as an n-tuple really depends both on i and n. So, we will drop the suffix n only when it is well understood from the context.

Remark A.2 (Simple projection arithmetics) *It is easy to check the following equalities:*

$$\mathrm{p}_1 \mathrm{p}_k^k M = \mathrm{p}_k^{k+1} M \ \ if \ k < n$$

$$\mathrm{p}_2 \mathrm{p}_k^k M = \mathrm{p}_{k+1}^{k+1} M \ \ if \ k < n$$

Lemma A.3 *Let M be a term in normal form and let $w : A_1 \times \cdots \times A_n$, where the A_i's have no occurrence of product or $\mathbf{T}$, be a variable not occurring free in it. Then $M[\overrightarrow{\mathrm{p}_i^n w}/\overrightarrow{y_i}]$ can only contain surjective pairing redexes. During any reduction to normal form of this term, the only created redexes are surjective pairing redexes.*

Proof. By inspection of the form of the redexes. $\square$

Theorem A.4 *Let $w : A_1 \times \cdots \times A_n$, where the A_i's have no occurrence of product or $\mathbf{T}$, be a variable not occurring free in a term Q in normal form. Then the following implications hold:*

1. if $Q[\overrightarrow{\mathrm{p}_i^n w}/\overrightarrow{y_i}] \xrightarrow{\beta^2 \eta^2 \pi *} \mathrm{p}_k^{n'} w$, then $Q = \begin{cases} y_k & \text{if } k < n' \\ y_n & \text{if } k = n' = n \\ \langle y_{n'}, \cdots, y_n \rangle & \text{otherwise} \end{cases}$

2. if $Q[\overrightarrow{\mathrm{p}_i^n w}/\overrightarrow{y_i}] \xrightarrow{\beta^2 \eta^2 \pi *} w$, then $Q = \langle y_1, \cdots, y_n \rangle$

Proof. By induction on the structure of Q. We notice here once and for all that both w and $\mathrm{p}_i^n w$ are in normal form w.r.t. the rewriting system we use for $\lambda^2 \beta \eta \pi *$, as their types do not contain occurrences of $*$, so not even *gentop* is an applicable reduction rule for them. This means that, in what follows, any time we find a term M equal to one of them, we can immediately assume that M reduces to one of them.

- **Q is a variable**

1. if Q is not one of the $y's$, then the claim holds vacuously; otherwise Q is some y_i and $Q[\overrightarrow{p_i^n w}/\overrightarrow{y_i}]$ is just $p_i^n w$, so that Q is as required;

2. holds vacuously

- **Q is an application $Q_1 Q_2$**
  The only new redexes created by the substitution are Surjective Pairing redexes, and their elimination cannot create any new β redexes (by Lemma A.3), so that there is no way to get rid of the outermost application and the claims hold vacuously.

- **Q is an abstraction $\lambda x.Q'$** The only way to get rid of the top-level abstraction in a reduction would be by means of an η reduction, but no such reductions are possible, due to Lemma A.3.

- **Q is a projection $p_1 Q'$**

  1. for $p_1 Q'[\overrightarrow{p_i^n w}/\overrightarrow{y_i}] \xrightarrow{\beta^2 \eta^2 \pi *} p_k^{n'} w$, we have two possibilities:

     - $p_k^{n'}$ is actually $p_1 p_k^n$ and $Q'[\overrightarrow{p_i^n w}/\overrightarrow{y_i}] \xrightarrow{\beta^2 \eta^2 \pi *} p_k^k w$. First notice that this implies that k is strictly smaller than n; otherwise the term $p_1 p_k^k w$ would not be well typed. Then, by inductive hypothesis 1, the term Q' is a pair $\langle y_k, \cdots, y_n \rangle$, and hence Q is not in normal form, contradicting the hypothesis of the theorem.

     - $p_k^{n'}$ does not start with a first projection, or $Q'[\overrightarrow{p_i^n w}/\overrightarrow{y_i}]$ does not reduce to $p_k^k w$. Then necessarily $Q'[\overrightarrow{p_i^n w}/\overrightarrow{y_i}] \xrightarrow{\beta^2 \eta^2 \pi *}$ to a pair $\langle p_k^{n'} w, \cdots \rangle$, in order for $p_1 Q'[\overrightarrow{p_i^n w}/\overrightarrow{y_i}]$ to reduce to $p_k^{n'} w$. But by Lemma A.3 in the reduction path there are only SP redexes, which can make pairs disappear but not create pairs. So Q' is already a product and Q is not in normal form, contradicting the hypothesis of the theorem.

  2. if $p_1 Q'[\overrightarrow{p_i^n w}/\overrightarrow{y_i}] \xrightarrow{\beta^2 \eta^2 \pi *} w$, then necessarily $Q'[\overrightarrow{p_i^n w}/\overrightarrow{y_i}] \xrightarrow{\beta^2 \eta^2 \pi *} \langle w, \cdots \rangle$, so that, similarly as for the previous point, Q' is already a pair and Q in not in normal form contradicting the hypothesis of the theorem.

- **Q is a projection $p_2 Q'$**

  1. for $p_2 Q'[\overrightarrow{p_i^n w}/\overrightarrow{y_i}] \xrightarrow{\beta^2 \eta^2 \pi *} p_k^{n'} w$, we have two possibilities:

- $p_k^{n'}$ is actually $p_{n'}^{n'}$ and $Q'[\overrightarrow{p_i^n w}/\overrightarrow{y_i}] \overset{\beta^2 \eta^2 \pi *}{\longrightarrow} p_{n'-1}^{n'-1}w$. First notice that $n'-1 < n' \leq n$. Then, by inductive hypothesis 1, the term Q' is a pair $\langle y_{n'-1}, \cdots, y_n \rangle$, and hence Q is not in normal form, contradicting the hypothesis of the theorem.

- $p_k^{n'}$ does not start with a second projection, or $Q'[\overrightarrow{p_i^n w}/\overrightarrow{y_i}]$ does not reduce to $p_{n'-1}^{n'-1}w$. Then necessarily $Q'[\overrightarrow{p_i^n w}/\overrightarrow{y_i}]$ $\overset{\beta^2 \eta^2 \pi *}{\longrightarrow}$ to a pair $\langle \cdots, p_{n'}^{n'}w \rangle$, in order for $p_2 Q'[\overrightarrow{p_i^n w}/\overrightarrow{y_i}]$ to reduce to $p_k^{n'}w$. But Lemma A.3 tells us that in the reduction path there are only η or SP redexes, which can make pairs disappear but not create pairs. So Q' is already a product and Q is not in normal form, contradicting the hypothesis of the theorem.

2. if $p_2 Q'[\overrightarrow{p_i^n w}/\overrightarrow{y_i}] \overset{\beta^2 \eta^2 \pi *}{\longrightarrow} w$, then necessarily $Q'[\overrightarrow{p_i^n w}/\overrightarrow{y_i}] \overset{\beta^2 \eta^2 \pi *}{\longrightarrow} \langle \cdots, w \rangle$, so that, similarly as for the previous point, Q' is already a pair and Q in not in normal form, contradicting the hypothesis of the theorem.

• **Q is a pair $\langle Q_1, Q_2 \rangle$**

1. for $\langle Q_1, Q_2 \rangle[\overrightarrow{p_i^n w}/\overrightarrow{y_i}] \overset{\beta^2 \eta^2 \pi *}{\longrightarrow} p_k^{n'}w$, we have two possibilities:

   - $k < n'$: then $p_k^{n'}w = p_1\underbrace{p_2 \cdots p_2}_{k-1} M = p_k^n w$, so the type of $p_k^{n'}w$ is A_k, which does not contain products and cannot be equal to a pair, whose type is a product.

   - $k = n'$: then $n' < n$ as $p_n^n w$ cannot be equal to a pair, because its type A_n contains no product. Then,

$$Q_1[\overrightarrow{p_i^n w}/\overrightarrow{y_i}] \overset{\beta^2 \eta^2 \pi *}{\longrightarrow} p_1 p_{n'}^{n'}w = p_{n'}^{n'+1}w = p_{n'}^n w,$$

and

$$Q_2[\overrightarrow{p_i^n w}/\overrightarrow{y_i}] \overset{\beta^2 \eta^2 \pi *}{\longrightarrow} p_2 p_{n'}^{n'}w = p_{n'+1}^{n'+1}w,$$

so by induction we have that

$$Q_1 = y_{n'}, \quad Q_2 = \langle y_{n'+1}, \cdots, y_n \rangle$$

and $Q = \langle y_{n'}, \cdots, y_n \rangle$ as required.

2. if $\langle Q_1, Q_2 \rangle[\overrightarrow{p_i^n w}/\overrightarrow{y_i}] \overset{\beta^2 \eta^2 \pi *}{\longrightarrow} w$, then we necessarily have the reductions $Q_1[\overrightarrow{p_i^n w}/\overrightarrow{y_i}] \overset{\beta^2 \eta^2 \pi *}{\longrightarrow} p_1 w = p_1^2 w$ and $Q_2[\overrightarrow{p_i^n w}/\overrightarrow{y_i}] \overset{\beta^2 \eta^2 \pi *}{\longrightarrow} p_2 w =$

$p_2^2 w$, so by induction hypothesis

$$Q_1 = y_1, \quad Q_2 = \langle y_2, \cdots, y_n \rangle$$

and

$$Q = \langle Q_1, Q_2 \rangle = \langle y_1, \cdots, y_n \rangle$$

as required.

- **Q is a second-order application $Q_1[B]$**
  There is no way to reduce Q_1 to a second-order abstraction by means of SP reductions (the only ones that are allowed by Lemma A.3), and it is not already one since Q is in normal form. So, there is no way to get rid of the second order application at the top level, and the two claims hold vacuously.

- **Q is a second-order abstraction $\lambda X.Q_1$**
  Again, there is no way to get rid of the top-level second-order abstraction, and the claims hold vacuously.

□

⚠ B   Technical lemmas

This section contains some technical lemmas and the proof of the main lemma 5.2.11. Before proceeding to the proof of the main lemma, let's study some properties of *DistInv*, in the following lemma, and then some connections between types and the structure of terms in the lemmas that follow.

Lemma B.1 *Let $DistInv(\vec{M}, \vec{V}, \vec{N}, \vec{W})$, let*

$$
\begin{aligned}
\vec{M}_{types} &= \{M_i \mid M_i \text{ is a type variable}\} \\
\vec{N}_{types} &= \{N_i \mid N_i \text{ is a type variable}\} \\
\vec{V}_{types} &= \{V_i \mid V_i \text{ is a type variable}\} \\
\vec{W}_{types} &= \{W_i \mid W_i \text{ is a type variable}\}
\end{aligned}
$$

and define

$$
\begin{aligned}
m_{type} &= |\vec{M}_{types}| & n_{type} &= |\vec{N}_{types}| \\
v_{type} &= |\vec{V}_{types}| & w_{type} &= |\vec{W}_{types}|
\end{aligned}
$$

then $m_{type} = n_{type} = v_{type} = w_{type}$.

Proof. First recall condition 3 of Definition 5.2.5. We have

$$M_i[\overrightarrow{N}/\overrightarrow{V}] = W_i \quad N_j[\overrightarrow{M}/\overrightarrow{W}] = V_j$$

for $\overrightarrow{V} = V_1, \cdots, V_m \subseteq FV(\overrightarrow{M})$ and $\overrightarrow{W} = W_1, \cdots, W_n \subseteq FV(\overrightarrow{N})$ sequences of variables, which can be either type variables X_i that are part of the list $\overrightarrow{M} = [M_1, \cdots, M_n]$ (resp. type variables Y_j from $\overrightarrow{N} = [N_1, \cdots, N_m]$), or term variables $x_i : A_i$ (resp. $y_i : B_i$).

Since in our calculus terms are not types, the members of the above equalities are either both types or both terms, so we can already establish that

$$m_{type} = w_{type} \quad \text{and} \quad n_{type} = v_{type}.$$

Now types do not contain terms either, so when we specialize condition 3 of Definition 5.2.5 to type expressions we get

$$\overrightarrow{M}_{types}[\overrightarrow{Y}/\overrightarrow{V}_{types}] = \overrightarrow{W}_{types} \quad \text{and} \quad \overrightarrow{N}_{types}[\overrightarrow{X}/\overrightarrow{W}_{types}] = \overrightarrow{V}_{types}.$$

Now, recall that the $\overrightarrow{M}_{types}$ and the $\overrightarrow{N}_{types}$ are simple type variables (Remark 5.2.6), and let's focus on the first equality. We know that no $X_i \in \overrightarrow{M}_{types}$ can be equal to a W_i, as W_i cannot be a free variable of $\overrightarrow{M}$, while X_i clearly is. This means that the substitution $[\overrightarrow{Y}/\overrightarrow{V}_{types}]$ must affect all the $\overrightarrow{M}_{types}$, and this can happen only if $\overrightarrow{V}_{types}$ includes $\overrightarrow{M}_{types}$, that is, if

$$m_{type} = |\overrightarrow{M}_{types}| \leq |\overrightarrow{V}_{types}| = v_{type} = n_{type}.$$

Analogously we get

$$n_{type} = |\overrightarrow{N}_{types}| \leq |\overrightarrow{W}_{types}| = w_{type} = m_{type}.$$

Hence $m_{type} = n_{type} = v_{type} = w_{type}$. $\square$

Lemma B.2 *Given $r\overrightarrow{P}: C$ where r is a variable and C a type containing an occurrence of a product type $A \times B$ then either $r{:}E$ and $D \times H$ occurs in E for some D, H, or some $P_i = [E]$ and $D \times H$ occurs in E for some D, H.*

Proof. By induction on the length n of $\overrightarrow{P}$.

Base: for $n = 0$ then $r : C$, which contains the product type $A \times B$ by hypothesis.

Inductive step: the lemma holds for $n \leq k$, so we prove it for $n = k + 1$. By cases on P_{k+1}:

- it is a term. Then $(r\overrightarrow{P_{1\ldots k}})P_{k+1} : C$ and $(r\overrightarrow{P_{1\ldots k}}) : Type(P_{k+1}) \to C$ so we can apply the induction hypothesis and get either that $r : E$ and $D \times H$ occurs in E for some D, H or that some $P_i = [E]$ and $D \times H$ occurs in E for some D, H for $i \leq k$, and hence for $i \leq k+1$ too.

- it is a type $[F]$. Then $(r\overrightarrow{P_{1\ldots k}}) : \forall X.T$ with $T[F/X] = C$. Now we have again two cases: either T does not contain any products, but then F must, or T contains products and we can apply induction as in the previous case.

□

Lemma B.3 *Given* $r\overrightarrow{P} : C$ *where* r *is a variable and* C *is a type containing an occurrence of* **T** *then either* $r : D$ *and* **T** *occurs in* D *or some* $P_i = [D]$, *and* **T** *occurs in* D.

Proof. By induction on the length n of $\overrightarrow{P}$.

- Base: for $n = 0$ then $r : C$ and C contains **T** by hypothesis.

- Inductive step: let the lemma be true for $n \leq k$ and prove it for $n = k+1$.

By cases on P_{k+1}:

  - it is a term. Then $(r\overrightarrow{P_{1\ldots k}}) : Type(P_{k+1}) \to C$, so we can apply the induction hypothesis and get either that $r : D$ and **T** occurs in D or that some $P_i = [D]$ and **T** occurs in D for $i \leq k$ and hence for $i \leq k+1$ too.

  - it is a type $[E]$. Then $(r\overrightarrow{P_{1\ldots k}}) : \forall X.T$ with $T[E/X] = C$. Now we have again two cases: either T does not contain **T**, but then E must, or T contains **T** and we can apply induction as in the previous case.

□

Lemma B.4 *Let* $z{:}A_0 \vdash (z\ \overrightarrow{P_{1\ldots n}}){:}\ C$. *If no product or* **T** *appears in the types* A_0, C *and in the* P_i's *that are types, then no product or* **T** *appears in the types of the* P_i's *that are terms either.*

Proof. By induction on n.
Base: $n = 0$ trivial.
Inductive step: let the lemma be true for $n \leq k$ and prove it for $n = k+1$
By cases on P_{k+1}:

- it is a term.

  By cases on the type E of P_{k+1}:

  - it has no occurrence of products or $\mathbf{T}$: then the type of $(r\overrightarrow{P_{1\ldots k}})$ $= E \to C$ has no occurrence of products or $\mathbf{T}$ too, so we can apply the induction hypothesis.

  - it has occurrences of products or $\mathbf{T}$; but this is impossible as $(r\overrightarrow{P_{1\ldots k}}) : E \to C$ contains products or $\mathbf{T}$, and so by Lemma B.2 or Lemma B.3 either r has a type containing products or $\mathbf{T}$ or some P_i's that are a type must contain a product or $\mathbf{T}$, in contradiction with the hypothesis.

- it is a type $[E]$. Then $(r\overrightarrow{P_{1\ldots k}}) : \forall X.T$ and in $\forall X.T$ there are no occurrences of products or $\mathbf{T}$ (as there are not in E by hypothesis and $T[E/X] = C$). So we can apply the induction hypothesis.

□

Lemma 5.2.11 (Main Lemma)

Let $\overrightarrow{N}$, $\overrightarrow{M}$ (with $m = |\overrightarrow{N}|$ and $n = |\overrightarrow{M}|$) be sequences of terms and/or types and $\circ$, $\bullet$ be head-free substitutions such that $DistInv(\overrightarrow{M}, \overrightarrow{x}, \overrightarrow{N}, \overrightarrow{y})^{\circ\bullet}$, then the following hold:

1. *the substitutions $\bullet$ and $\circ$ are idempotent, i.e., $\bullet\bullet = \bullet$ and $\circ\circ = \circ$*

2. *when M_i and N_j are terms,*

$$M_i = \lambda\overrightarrow{v_i}.x_{\sigma(i)}\overrightarrow{P_i}, \quad N_j = \lambda\overrightarrow{u_j}.y_{\pi(j)}\overrightarrow{Q_j}$$

   where $\sigma{:}n \to m$, $\pi{:}m \to n$ are integer functions s.t. $\sigma(\pi(i))=i$ if M_i is a term and $\pi(\sigma(j))=j$ if N_j is a term

3. *$n = m$*

4. *every P_{i_k} (Q_{j_h}) that is a type expression is just a type variable*

5. *every variable free in P_{i_k} (Q_{j_h}) has no occurrence of $\times$ or $\mathbf{T}$ in the type expressions occurring in it*

6. *P_{i_k} (Q_{j_h}) has no occurrence of $\times$ or $\mathbf{T}$ in its type if P_{i_k} (Q_{j_h}) is a term*

7. *Define* $s_1 = | \overrightarrow{P}_i |, r_2 = | \overrightarrow{u}_{\sigma(i)} |, r_1 = | \overrightarrow{v}_i |, s_2 = | \overrightarrow{Q}_{\sigma(i)} |$. *Without loss of generality, suppose* $s_1 \geq r_2$. *Then* $r_1 \geq s_2$. *Furthermore,* $\overrightarrow{Q}_{\sigma(i)}$ *can be extended with a sequence of type and/or term variables* $[u''_1, \cdots, u''_{r_1-s_2}]$ *to a sequence* $\overrightarrow{Q'}_{\sigma(i)}$ *such that*

$DistInv(\overrightarrow{P}_i, \overrightarrow{v}_i, \overrightarrow{Q'}_{\sigma(i)}, \overrightarrow{u'}_{\sigma(i)})^{\bowtie}$ *holds, where* $\overrightarrow{u'}_{\sigma(i)} = \overrightarrow{u}_{\sigma(i)} \cup [u''_1, \cdots, u''_{r_1-s_2}]$ *and* ◁, ▷ *are suitable head-free substitutions.*

Proof. We will show properties 1 - 7 in order, but we factor out here a remark that is needed along most of the proof. We will very often make use of the Church-Rosser rewrite system associated to the calculus in order to exclude possible reductions. In all these cases the reduction (*gentop*) will not be possible, since in order to be applied to a term M, *gentop* requires that M has a type in $Iso(\mathbf{T})$, which will never be the case in the following since no $\mathbf{T}$ type will be involved and since any type in $Iso(\mathbf{T})$ contains at least one occurrence of $\mathbf{T}$. We will not state this fact explicitly all the time.

1. is trivial due to the requirements 1 and 2 on the domain and codomain of the head-free substitutions ∘ and • in the definition 5.2.10 of $DistInv(,,,)$.

2. since the M_i in $\overrightarrow{M}$ and the N_i in $\overrightarrow{N}$ are terms are in normal form, we know that

$$M_i = \lambda \overrightarrow{v_i}.R_i \overrightarrow{P_i}$$

and

$$N_i = \lambda \overrightarrow{u_i}.S_i \overrightarrow{Q_i},$$

where R_i, $\overrightarrow{P_i}$ (respectively S_i, $\overrightarrow{Q_i}$) are terms in normal form and R_i (respectively S_i) is not an abstraction nor an application (first- or second-order).

Now notice that R_i (respectively S_i)

- cannot be a pair $\langle R_{i_1}, R_{i_2} \rangle$, as otherwise, for typing reasons, $M_i = \lambda \overrightarrow{v_i}.\langle R_{i_1}, R_{i_2} \rangle$ has type $A \to \cdots \to B \times C$ that is not an arrow-only type;

- cannot be $*:\mathbf{T}$, as otherwise the type of the term would contain $\mathbf{T}$;

- cannot be a bound variable, as otherwise $M_i^{\bullet}[\overrightarrow{N}^{\circ} / \overrightarrow{x}]$ could not reduce to the free variable y_i. Indeed, if it is bound, then $M_i^{\bullet}[\overrightarrow{N}^{\circ} / \overrightarrow{x}]$ is $\lambda \overrightarrow{v_i}.R_i \overrightarrow{P^{\bullet}}_i[\overrightarrow{N}^{\circ} / \overrightarrow{x}]$, since a substitution affects only free variables. So R_i is still bound. Now, y_i is in normal form, and the equality $M_i^\bullet[\overrightarrow{N}^\circ/\overrightarrow{x}] = y_i$ can be turned into a reduction sequence $M_i^\bullet[\overrightarrow{N}^\circ/\overrightarrow{x}] \overset{*}{\to} y_i$ by the Church-Rosser Property. But no reduction rule can make disappear from a term in head normal-form a bound variable that is in the head position.

We want to show that R_i cannot be a projection $\mathrm{p}_k Q$ either. Supposing that R_i is $\mathrm{p}_k Q$ $(k = 1, 2)$, then

$$M_i = \lambda \overrightarrow{v_i}.\mathrm{p}_k Q P_{i_1} \cdots P_{i_l}$$

with Q in normal form. More precisely, $Q = O\overrightarrow{R_i}$, where O is in normal form and is not an abstraction or an application.[3] Again, O cannot be a pair $\langle O_1, O_2 \rangle$ (as otherwise $Q = \langle O_1, O_2 \rangle$ and the subterm $\mathrm{p}_k Q = \mathrm{p}_k \langle O_1, O_2 \rangle$ of M_i would be a π redex, while we know that M_i is in n.f.) nor a bound variable (for the same reasons shown above for R_i). This argument can be iterated to show that

$$M_i = \lambda \overrightarrow{v_i}.\mathrm{p}_{k_1}(\cdots(\mathrm{p}_{k_z}(r\overrightarrow{O}_{z+1})\overrightarrow{O}_z)\cdots\overrightarrow{O}_1)P_{i_2}\cdots P_{i_l}$$

with r a free variable.

With the same argument we show that

$$N_i = \lambda \overrightarrow{u_i}.\mathrm{p}_{s_1}(\cdots(\mathrm{p}_{s_w}(s\overrightarrow{U}_{w+1})\overrightarrow{U}_w)\cdots\overrightarrow{U}_1)Q_{i_2}\cdots Q_{i_h}.$$

Now we can use the property $\overrightarrow{M}_i^\bullet[\overrightarrow{N}^\circ/\overrightarrow{x}] = y_i$ to show that $r \in \overrightarrow{x}$ considering the following cases:

(a) If $\bullet$ does not affect r, then r must be one of the $\overrightarrow{x}$; otherwise it is impossible to reduce the sequence of projections in front of r or (if the sequence is empty) to reduce $\overrightarrow{M}_i^\bullet[\overrightarrow{N}^\circ/\overrightarrow{x}]$ to a term not containing r, as r is the head variable in $\overrightarrow{M}_i^\bullet[\overrightarrow{N}^\circ/\overrightarrow{x}]$ too.

(b) Otherwise, notice that $\bullet$ is a head-free substitution and cannot, by hypothesis, generate any y_i, so that again $\overrightarrow{M}_i^\bullet[\overrightarrow{N}^\circ/\overrightarrow{x}]$ has a head free variable w that is not any y_i and cannot be erased by reductions, no matter if the sequence of projections in front of it is empty or not.

---

[3] If it is an abstraction, then the term is not in normal form, while if it is an application $O_1 O_2$ we would take O_1 instead. Notice also that Q itself is not an abstraction for typing reasons.

So r must be some $x_{\sigma(i)}$ for some integer function $\sigma : n \to m$. Analogously, s must be some $y_{\pi(i)}$ for some integer function $\pi : m \to n$. This means that

$$\overset{\bullet}{\vec{M}}_i[\overset{\circ}{\vec{N}}/\vec{x}] = \lambda \vec{v_i}.\mathrm{p}_{k_1}(\cdots(\mathrm{p}_{k_z}(\mathrm{N}^{\circ}_{\sigma(i)}\overset{\triangleright}{\vec{\mathcal{O}}}_{z+1})\overset{\triangleright}{\vec{\mathcal{O}}}_z)\cdots\overset{\triangleright}{\vec{\mathcal{O}}}_1)\overset{\triangleright}{\vec{P}}_{i_2\cdots i_l}$$

where $\triangleright$ is the substitution $\bullet[\overset{\circ}{\vec{N}}/\vec{x}]$. But since

$$N_{\sigma(i)} = \lambda\overrightarrow{u_{\sigma(i)}}.\mathrm{p}_{s_1}(\cdots(\mathrm{p}_{s_w}(y_{\pi(\sigma(i))}\vec{U}_{w+1})\vec{U}_w)\cdots\vec{U}_1)\vec{Q}_{\sigma(i)_2\cdots\sigma(i)_h}$$

then if $|\overrightarrow{u_{\sigma(i)}}| \leq |\overset{\triangleright}{\vec{\mathcal{O}}}_{z+1}|$ we have

$$\overset{\bullet}{\vec{M}}_i[\overset{\circ}{\vec{N}}/\vec{x}] =$$
$$= \lambda\vec{v_i}.\mathrm{p}_{k_1}(\cdots$$
$$(\mathrm{p}_{k_z}(((\mathrm{p}_{s_1}(\cdots(\mathrm{p}_{s_w}(y_{\pi(\sigma(i))}\overset{\circ\circ}{\vec{U}}_{w+1})\overset{\circ\circ}{\vec{U}}_w)\cdots\overset{\circ\circ}{\vec{U}}_1))$$
$$\overset{\circ\circ}{\vec{Q}}_{\sigma(i)_2\cdots\sigma(i)_h})\overset{\triangleright}{\vec{\mathcal{O}'}}_{z+1})\overset{\triangleright}{\vec{\mathcal{O}}}_z)$$
$$\cdots\overset{\triangleright}{\vec{\mathcal{O}}}_1)\overset{\triangleright}{\vec{P}}_{i_2\cdots i_l}$$

(where $\overset{\triangleright}{\vec{\mathcal{O}'}}_{z+1}$ is what is left of $\overset{\triangleright}{\vec{\mathcal{O}}}_{z+1}$ after the β reductions on $(\overset{\circ}{\vec{N}}_{\sigma(i)}$ $\overset{\triangleright}{\vec{\mathcal{O}}}_{z+1})$ and $\diamond$ is the substitution $[\overset{\triangleright}{\vec{\mathcal{O}}}_{z+1}/\overrightarrow{u_{\sigma(i)}}]$). Otherwise

$$\overset{\bullet}{\vec{M}}_i[\overset{\circ}{\vec{N}}/\vec{x}] =$$
$$= \lambda\vec{v_i}.\mathrm{p}_{k_1}(\cdots$$
$$(\mathrm{p}_{k_z}(\lambda\vec{u'}_{\sigma(i)}.((\mathrm{p}_{s_1}(\cdots(\mathrm{p}_{s_w}(y_{\pi(\sigma(i))}\overset{\circ\circ}{\vec{U}}_{w+1})\overset{\circ\circ}{\vec{U}}_w)\cdots\overset{\circ\circ}{\vec{U}}_1))$$
$$\overset{\circ\circ}{\vec{Q}}_{\sigma(i)_2\cdots\sigma(i)_h}))\overset{\triangleright}{\vec{\mathcal{O}}}_z)$$
$$\cdots\overset{\triangleright}{\vec{\mathcal{O}}}_1)\overset{\triangleright}{\vec{P}}_{i_2\cdots i_l}$$

(where $\vec{u'}_{\sigma(i)}$ is what is left of $\vec{u}_{\sigma(i)}$ after the β reductions on $(\overset{\circ}{\vec{N}}_{\sigma(i)}$ $\overset{\triangleright}{\vec{\mathcal{O}}}_{z+1})$ [4] and $\diamond$ is the substitution $[\overset{\triangleright}{\vec{\mathcal{O}}}_{z+1}/\overrightarrow{u_{\sigma(i)}}]$).

In any case, these terms must be equal to y_i, which is a normal form, and equality can be turned into reduction due to the Church-Rosser Property. Now, this reduction is possible iff $z = w = 0$ and

---

[4] Actually, for typing reasons, this sequence must be empty. We have a projection p_{k_z} applied to this term that cannot have, then, a functional type!

$y_{\pi(\sigma(i))} = y_i$, i.e., $\pi(\sigma(i)) = i$. There is no reduction rule that allows us to get rid of the sequence of projections because they are blocked by the variable $y_{\pi(\sigma(i))}$ (that is free and is already in normal form). So r_i must be $x_{s(i)}$ and $\pi(\sigma(i)) = i$ for i s.t. M_i is a term. Analogously we can proceed for $\overrightarrow{N}$ and get that s_j must be $y_{\pi(j)}$ and $\sigma(\pi(j)) = j$ for j s.t. N_j is a term.

3. Since we already know that the number of type variables in $\overrightarrow{M}$, $\overrightarrow{N}$, $\overrightarrow{x}$ and $\overrightarrow{y}$ is equal (Lemma B.1) it is possible to extend the previous functions π and σ in order to get $\pi(\sigma(i)) = i$ and $\sigma(\pi(j)) = j$ for all i and j. Just let σ map indexes of different type variables in $\overrightarrow{M}$ to indexes of different type variables in $\overrightarrow{x}$ and let π be the inverse of σ on the indexes of type variable of $\overrightarrow{N}$.

Namely, from Lemma B.1 we obtain also that the type variables in both $\overrightarrow{M}$ and $\overrightarrow{N}$ are just a permutation of each other, so that there exists $\sigma_{type}:n \to m$, $\pi_{type}:m \to n$ s.t. $\sigma_{type}(\pi_{type}(i)) = i$ if M_i is a type variable and $\pi_{type}(\sigma_{type}(j)) = j$ when N_j is a type variable. Hence we can define $\sigma' : n \to m$, $\pi' : m \to n$ as

- $\sigma'(i) = \sigma(i)$ if M_i is a term

- $\sigma'(i) = \sigma_{type}(i)$ if M_i is a type variable

- $\pi'(i) = \pi(j)$ if N_j is a term

- $\pi'(i) = \pi_{type}(j)$ if N_j is a type variable

with the property that $\pi'(\sigma'(i)) = i$ and $\sigma'(\pi'(j)) = j$ for all i and j, since every M_i and N_j is either a term or a type variable.

Due to well-known properties of permutations, this entails $n = m$, and furthermore π' and σ' are permutations that are each other's inverses. This says, besides, that the number of terms is the same in both sequences.

4. Let's consider again $\overrightarrow{M}_i^{\bullet}[\overrightarrow{N}^{\circ}/\overrightarrow{x}]$. Now we know from 3 and the proof of 2 that

$$
\begin{aligned}
\overrightarrow{M}_i^{\bullet}[\overrightarrow{N}^{\circ}/\overrightarrow{x}] = {} \\
= {} & \lambda\overrightarrow{v_i}.\mathrm{N}^{\circ}_{\sigma(i)}\mathrm{P}^{\triangleright}_{i_2}\cdots\mathrm{P}^{\triangleright}_{i_l} \text{ (where } \triangleright \text{ is } \bullet[\overrightarrow{N}^{\circ}/\overrightarrow{x}]) \\
= {} & \lambda\overrightarrow{v_i}.(\lambda\overrightarrow{u_{\sigma(i)}}.y_{\pi(\sigma(i))}\overrightarrow{Q}^{\circ}_{\sigma(i)})\mathrm{P}^{\triangleright}_{i_2}\cdots\mathrm{P}^{\triangleright}_{i_l} \\
= {} & \lambda\overrightarrow{v_i}.(\lambda\overrightarrow{u_{\sigma(i)}}.y_i\overrightarrow{Q}^{\circ}_{\sigma(i)})\mathrm{P}^{\triangleright}_{i_2}\cdots\mathrm{P}^{\triangleright}_{i_l} \\
= {} &
\begin{cases}
\lambda\overrightarrow{v_i}.(\lambda\overrightarrow{u'_{\sigma(i)}}.y_i\overrightarrow{Q}^{\circ}_{\sigma(i)})[\mathrm{P}^{\triangleright}_{i_2}\cdots\mathrm{P}^{\triangleright}_{i_l}/\overrightarrow{u_{\sigma(i)} - u'_{\sigma(i)}}] \\
\quad \text{if } |\overrightarrow{u_{\sigma(i)}}| > l-1 \\
\lambda\overrightarrow{v_i}.((y_i\overrightarrow{Q}^{\circ}_{\sigma(i)})[\mathrm{P}^{\triangleright}_{i_2}\cdots\mathrm{P}^{\triangleright}_{i_{l'}}/\overrightarrow{u_{\sigma(i)}}])\mathrm{P}^{\triangleright}_{i_{l'+1}}\cdots\mathrm{P}^{\triangleright}_{i_l} \\
\quad \text{otherwise.}
\end{cases}
\end{aligned}
$$

Some care is needed in checking the last equality: in the case $|\overrightarrow{u}_{\sigma(i)}| > l-1$, we named $\overrightarrow{u'}_{\sigma(i)}$ the abstracted variables that are left after the β reductions, and the notation $[\mathrm{P}^{\triangleright}_{i_2}\cdots\mathrm{P}^{\triangleright}_{i_l}/\overrightarrow{u_{\sigma(i)} - u'_{\sigma(i)}}]$ is a shorthand to indicate that the substitution is performed on the first $l-1$ variables of $\overrightarrow{u_{\sigma(i)}}$.

The only way to reduce both expressions to y_i is a series of η reductions. This means that every Q_i type expression must be a type variable. The same can be shown for P_i considering $\overrightarrow{N}_i^{\circ}[\overrightarrow{M}^{\bullet}/\overrightarrow{y}]$ instead.

5. Any free variable in any P_i either is free in $\overrightarrow{M}$ (and then the claim holds by hypothesis) or is one of the $\overrightarrow{v_i}$. In the second case, notice that the type of the $\overrightarrow{M}$ contains as type subexpressions the type of the $\overrightarrow{v_i}$, so that it cannot contain any product or $\mathbf{T}$ either.

6. By Lemma B.4 (the lemma on type of sequences not including complex type expressions) and properties 4 and 5 (recall that $\mathbf{T}$ is a type constant, not a variable).

7. W.l.o.g., let $s_1(=|\overrightarrow{P}_i|) \geq r_2(=|\overrightarrow{u}_{\sigma(i)}|)$. Notice that this implies $r_1(=|\overrightarrow{v}_i|) \geq s_2(=|\overrightarrow{Q}_{\sigma(i)}|)$, as by inspecting both cases in the proof of 4 we get $r_1 + r_2 = s_1 + s_2$ (due to the η reductions that must occur). Then $\overrightarrow{Q}_{\sigma(i)}$ can be extended with a sequence of type and/or term variables $[u''_1, \cdots, u''_{r_1 - s_2}]$ to a $\overrightarrow{Q'}_{\sigma(i)}$ so that (we will drop the

suffix $\sigma(i)$ now for clarity)

$$(\vec{Q}^{\circ}_{1\cdots s_2}[\vec{P}^{\triangleright}_{1\cdots r_2}/\vec{u}_{1\cdots r_2}])\vec{P}^{\triangleright}_{r_2+1\cdots s_1} =$$

$$= (\vec{Q}^{\circ}_{1\cdots s_2}[\vec{P}^{\triangleright}_{1\cdots r_2}/\vec{u}_{1\cdots r_2}])\vec{u}''_{1\cdots r_1-s_2}[\vec{P}^{\triangleright}_{r_2+1\cdots s_1}/\vec{u}''_{1\cdots r_1-s_2}]$$

$$= \vec{Q'}^{\circ}[\vec{P}^{\triangleright}_{1\cdots r_2}/\vec{u}_{1\cdots r_2}, \cdots, \vec{P}^{\triangleright}_{r_2+1\cdots s_1}/\vec{u}''_{1\cdots r_1-s_2}]$$

$$= \vec{Q'}^{\circ}[\vec{P}^{\triangleright}_{1\cdots s_1}/\vec{u}_{1\cdots s_1}]$$

$$= v_1, \cdots, v_{r_1}$$

while

$$\vec{P}^{\bullet}[\vec{Q}^{\triangleleft}_{1\cdots s_2}u''_1, \cdots, u''_{r_1-s_2}/\vec{v}_{1\cdots r_1}] = u_1, \cdots, u_{r_2}, u'_1, \cdots, u'_{r_1-s_2}$$

where $\triangleleft$ is the substitution $\circ[\vec{M}^{\bullet}/\vec{y}]$, which plays the symmetric role of $\triangleright$ on the $\vec{Q}$.

Now notice that $[\vec{M}^{\bullet}/\vec{y}]$ does not affect any u's and does not create any v's, since by the variable convention they are not free in the M's.

Symmetrically, $[\vec{N}^{\circ}/\vec{x}]$ does not affect the v's and does not create any u's, so we get also:

$$\vec{Q'}^{\triangleleft}[\vec{P}^{\triangleright}_{1\cdots s_1}/\vec{u}_{1\cdots s_1}] = \vec{Q'}^{\circ}[\vec{M}^{\bullet}/\vec{y}][\vec{P}^{\triangleright}_{1\cdots s_1}/\vec{u}_{1\cdots s_1}]$$

$$= v_1, \cdots, v_{r_1}$$

and

$$\vec{P}^{\triangleright}[\vec{Q}^{\triangleleft}_{1\cdots s_2}u''_1, \cdots, u''_{r_1-s_2}/\vec{v}_{1\cdots r_1}]$$

$$= \vec{P}^{\bullet}[\vec{N}^{\circ}/\vec{x}][\vec{Q}^{\triangleleft}_{1\cdots s_2}u''_1, \cdots, u''_{r_1-s_2}/\vec{v}_{1\cdots r_1}]$$

$$= u_1, \cdots, u_{r_2}, u''_1, \cdots, u''_{r_1-s_2}$$

$$= u'_1, \cdots, u'_{s_1}$$

so that $DistInv(\vec{P_i}, \vec{v_i}, \vec{Q'}_{\sigma(i)}, \vec{u}_{\sigma(i)})^{\triangleright\triangleleft}$.

This concludes the proof of the main lemma. $\square$

C   Miscellanea

Proposition 5.1.2 *Each type has a unique type normal form in* $\mathcal{R}^2$.

Proof. Here is a direct proof of the proposition.

We show that $\mathcal{R}^2$ is a strongly normalizing Church-Rosser rewriting system. Since it is straightforward to show that the system is weakly Church-Rosser, it is enough to show SN (as, due to the well-known Newman's lemma, WCR + SN $\Rightarrow$ CR).

We will show strong normalization by exhibiting a measure that strictly decreases each time one of the rewriting rules is applied to a formula. Let h be a complexity measure on formulae defined as follows:

$$
\begin{aligned}
h(A) &= 3 \text{ if } A \text{ is atomic} \\
h(A \times B) &= h(A) * h(B)^2 + 1 \\
h(A \to B) &= h(B)^{h(A)} \\
h(\forall X.A) &= 2^{h(A)}.
\end{aligned}
$$

It is obvious that this measure is an integer always greater than 2. Then it is easy to show by induction that $h(C[A]) > h(C[A'])$ if $h(A) > h(A')$, where $C[\ ]$ is an arbitrary context. To show that every rewriting decreases h, it suffices now to show that every reduction rule in $\mathcal{R}^2$ strictly decreases this measure. Since 5, 6, 7, 8 and 9 trivially decrease h, we will focus only on 1, 2, 3 and 4.

$$
\begin{aligned}
h(A \times (B \times C)) &= h(A) * (h(B \times C))^2 + 1 \\
&= h(A) * \left(h(B) * h(C)^2 + 1\right)^2 + 1 \\
&> h(A) * h(B)^2 * h(C)^4 + 1 \\
&= \left(h(A) * h(B)^2 * h(C)^2\right) * h(C)^2 + 1 \\
&> h(A) * h(B)^2 * h(C)^2 + h(C)^2 + 1 \\
&\qquad \text{as } h(C)^2 > 2 \text{ and } h(A) * h(B)^2 * h(C)^2 > 2 \\
&= \left(h(A) * h(B)^2 + 1\right) * h(C)^2 + 1 \\
&= h(A \times B) * h(C)^2 + 1 \\
&= h((A \times B) \times C).
\end{aligned}
$$

Similarly,

$$
\begin{aligned}
h((A \to B) \times C) &= h(C)^{(h(A)*h(B)^2+1)} \\
&> h(C)^{(h(A)*h(B))}
\end{aligned}
$$

$$= \quad h(A \to (B \to C))$$

$$
\begin{aligned}
h(A \to (B \times C)) \quad &= \quad (h(B) * h(C)^2 + 1)^{h(A)} \\
&> \quad (h(B) * h(C)^2)^{h(A)} + 1 \\
&\qquad \text{as } h(A) > 2 \text{ and } h(B) * h(C)^2 > 2 \\
&= \quad h(B)^{h(A)} * h(C)^{2*h(A)} + 1 \\
&= \quad h(B)^{h(A)} * \left(h(C)^{h(A)}\right)^2 + 1 \\
&= \quad h((A \to B) \times (A \to C))
\end{aligned}
$$

$$
\begin{aligned}
h(\forall X.A \times B) \quad &= \quad 2^{(h(A)*h(B)^2+1)} \\
&= \quad 2 * 2^{h(A)*h(B)^2} \\
&= \quad 2 * 2^{h(A)} * 2^{h(B)^2} \\
&\geq \quad 2^{h(A)} * 2^{h(B)^2} + 1 \\
&\qquad \text{as } 2^{h(A)} * 2^{h(B)^2} \geq 1 \\
&> \quad 2^{h(A)} * 2^{2*h(B)} + 1 \\
&\qquad \text{as } h(B)^2 > 2 * h(B) \text{ since } h(B) > 2 \\
&= \quad 2^{h(A)} * \left(2^{h(B)}\right)^2 + 1 \\
&= \quad h(\forall X.A \times \forall X.B)
\end{aligned}
$$

□

Proposition C.1 *If A is a type not containing* **T**, *then there is no invertible term M of type $A \to$ **T**. Hence* **T** *is not definably isomorphic to any such type A.*

Proof. We can assume M is in normal form with respect to the confluent notion of reduction introduced in 2.3.2. Due to its type, M is $\lambda x : A.*$, and we show that no term N in normal form can be its inverse, considering the possible normal forms of N that would be compatible with its type. There are five cases:

- N is not $x : \mathbf{T} \to A$, as

$$
\begin{aligned}
x \circ M \quad &= \quad \lambda w : A.(x(Mw)) \\
&= \quad \lambda w : A.(x((\lambda x : A.*)w)) \\
&= \quad \lambda w : A.(x*),
\end{aligned}
$$

which is in normal form and is not the identity of type A.

- N is not $\lambda x : \mathbf{T}.P$, as

$$
\begin{aligned}
(\lambda x : \mathbf{T}.P) \circ M &= \lambda w : A.((\lambda x : \mathbf{T}.P)(Mw)) \\
&= \lambda w : A.((\lambda x : \mathbf{T}.P)((\lambda x : A.*)w)) \\
&= \lambda w : A.((\lambda x : \mathbf{T}.P)*) \\
&= \lambda w : A.(P[*/x]) \\
&= \lambda w : A.P
\end{aligned}
$$

(because $\lambda x : \mathbf{T}.P$ is in n.f., hence $x \notin FV(P)$,

as otherwise a reduction $x : \mathbf{T} \xrightarrow{\mathbf{T}} *$ could apply).

Since P is normal, $\lambda w : A.P$ is normal too, and due to the variable convention P cannot be w, so that $\lambda w : A.P$ is not the identity of type A.

- N is not (PQ), as

$$
\begin{aligned}
(PQ) \circ M &= \lambda w : A.((PQ)(Mw)) \\
&= \lambda w : A.((PQ)((\lambda x : A.*)w)) \\
&= \lambda w : A.((PQ)*)
\end{aligned}
$$

which is in normal form, as (PQ) is and $((PQ)*)$ is not a *gentop* redex because its type is A, which contains no occurrence of $\mathbf{T}$, and is not the identity of type A.

- N is not $\lambda X.P$ for typing reasons.

- N is not $P[B]$ for any type B, as

$$
\begin{aligned}
(P[B]) \circ M &= \lambda w : A.((P[B])(Mw)) \\
&= \lambda w : A.((P[B])((\lambda x : A.*)w)) \\
&= \lambda w : A.((P[B])*)
\end{aligned}
$$

which is in normal form, as $P[B]$ is and $((P[B])*)$ is not a *gentop* redex as its type is A, which contains no occurrence of $\mathbf{T}$, and is not the identity of type A.

It follows that M as in the hypothesis has no inverse. $\square$

Chapter 6

Isomorphisms for ML

This chapter is a first attempt to a formal foundation for the theory of isomorphisms of types in the framework of type-assignment calculi similar to Milner's ML [Mil78, MT91, CH88, MTH90]. The usual notion of isomorphism is not adequate for type assignment, because there a term M is no longer assigned a unique type, so that writing $M : A \to B$ is not only ambiguous but also incorrect.

We introduce an adequate new notion, and we proceed to investigate the isomorphisms of types: a new one shows up, called here (**Split**), which is not provable in the explicitly typed versions of the language. Such new isomorphism allows the design of a search algorithm which is not only correct but also more efficient than the previously proposed ones.

These results allow to develop a complete practical implementation of a library search system for both the CAML and CamlLight functional languages based on equality of types modulo isomorphisms. The algorithm is fully described for the case of the CAML system, but it can easily be rewritten for other dialects of ML or different functional languages. If you are just looking for a guide to the implementation of a library search system, then Sections 6.4 and 6.4.1 are all you need to see.

On the other side, the (**Split**) isomorphism has something peculiar to it that suggests its use in extending the traditional ML type inference algorithm. Such extension is very simple to implement (it has been developed as a patch to Caml-Light V0.4), and we consider it more a completion than an extension of the traditional system, since it does not introduce radically new typing mechanisms but allows us to give the same type to equivalent expressions that are treated differently by the original type inference mechanism.

6.1   Introduction

The known results on isomorphisms of types for explicitly typed languages are not adequate to handle languages where the *let* polymorphic construct is allowed. We study in Section 6.2 the problem of valid isomorphisms *in the presence* of such a polymorphic construct, and we uncover an isomorphism that does not hold for explicitly typed second-order typed λ-calculus. This new isomorphism allows us to provide a complete and decidable characterization for isomorphism of the core ML-like languages (Section 6.3). Furthermore, it makes it possible to design an efficient decision procedure for the complete theory of isomorphic types, which can be effectively used in an actual library search system. We provide an extensive description of this procedure, along with the actual code for the CAML dialect of ML, in Section 6.4.1. This new isomorphism can also be used to extend in a conservative way the usual ML type-inference algorithm, as we will see in the second part of the chapter, starting from Section 6.5.

6.2 Isomorphisms of types in ML-style languages

In [Rit91] and [Rit90a], Rittri uses the theory $Th^1_{\times \mathbf{T}}$ to develop a library search system for strongly typed functional languages in the ML family. Languages of the ML family are equipped with the so-called "implicit type polymorphism", a brand of type polymorphism that essentially allows us to give the user the safety of a strongly typed world without the burden of mandatory type declarations; the user writes type-free programs and the compiler "infers" a type for them by filling in all the type information.

The inference problem is easily decidable in the case of monomorphic languages, like the simply typed λ-calculus, (see [Hin69], [Mil78]), while we do not know how to deal with it for calculi with the full power of second-order quantification over types, like second-order typed λ-calculus.

It is a common idea (but we will shortly see how it is not a very correct one) that ML-style languages lie somewhere in between these two extremes, since any user-defined function is given a type that can be more than monomorphic, but not fully second-order polymorphic. These types are either monomorphic types (known as *monotypes* and denoted by τ below) or the so-called type schemas (denoted by σ below).

Definition 6.2.1 *ML types are the* closed *types generated by the following grammar (At is a collection of atomic types):*

type schemas $\sigma ::= \tau \mid \forall X.\sigma$ *(if X is free in σ)*
monotypes $\tau ::= At \mid X \mid \tau \to \tau \mid \tau \times \tau$

Type-schemas are essentially types where every type variable is bound by a quantifier that can appear only as an outermost constructor of the type (and not inside $\to$, $\times$ or other type-constructors).

If we follow the common intuition that ML is somewhere in between simple typed λ-calculus and second-order λ-calculus, it is easy to conjecture that the valid isomorphisms of type schemas are axiomatized by a theory Th^{ML} that includes $Th^1_{\times \mathbf{T}}$ and is included in $Th^2_{\times \mathbf{T}}$.

Then, noticing that Axioms 10, 11 and 12 involve second-order types that are not type schemas, it seems reasonable that Th^{ML} be just $Th^2_{\times \mathbf{T}}$ less these three axioms. So the naive approach to deciding equality of type schemas $\sigma_1 = \forall \overrightarrow{X}.\tau_1$ and $\sigma_2 = \forall \overrightarrow{Y}.\tau_2$ would be to check if there is a way of substituting in some order the variables $\overrightarrow{X}$ with $\overrightarrow{Y}$ in τ_1 such that for the resulting type τ_1' the theory $Th_{\times \mathbf{T}}^1$ proves $\tau_1' = \tau_2$. We say naive, because in principle the restriction of $Th_{\times \mathbf{T}}^2$ to ML types is not necessarily axiomatized by the restriction to ML types of the axiomatic presentation $Th_{\times \mathbf{T}}^2$ that we have chosen for this equality relation. Even worse, the techniques used to show completeness for $Th_{\times \mathbf{T}}^2$ on second-order types rely in an essential way on the fact that the language considered there is explicitly typed, while ML-style languages are type-assignment systems equipped with a *let* construct whose typing rules have no immediate counterpart in the explicitly typed calculi. So we could expect to find some isomorphism that is not axiomatized even in the full theory $Th_{\times \mathbf{T}}^2$.

Rittri's system (see [Rit91]), based on the well-known soundness of $Th_{\times \mathbf{T}}^1$ for monomorphic languages, implements precisely the procedure sketched above, and is *sound* for isomorphisms.in ML, but to handle the *completeness* problem in ML we have to face the problem of valid type schema isomorphisms in its own right. It turns out that we are in for some surprises here, but first of all, let's set up the right formalism for type-assignment systems.

6.2.1   A formal setting for valid isomorphisms in ML-like languages

First we will give the basic typing rules for ML-like languages. Instead of using the original presentation of ML proposed in [DM82] and shown in Table 1.1, we prefer to work with the variant given by [C$^+$86a], which has the advantage of being syntax-directed and of producing normalized typing derivations:

Definition 6.2.2 (type assignment) *We write* $\Gamma \vdash M : A$ *if* M *can be assigned type* A *in the type-assignment system given in Table 6.1.*

Remark 6.2.3 *Notice that the (LET) rule gets priority on the ordinary (APP) rule, that is to say, if we have to type an application* (MN)*, we use rule (LET) if* M *is a lambda abstraction and rule (APP) otherwise. Notice also that the (GEN) rule is actually usable only as the last rule of a typing judgment: all other rules accept only terms with monotypes. This rule is here only to allow the inference of a polymorphic type for a term, as is done in the usual implementations of the language.*

In the traditional presentations, one avoids the overlapping of rules (LET) and (APP) by introducing the notation *let* $x = e'$ *in* e for $(\lambda x.e)e'$, but it is important to remark that this new notation is just syntactic sugar.

$(UNIT)$ $\quad \Gamma \vdash () : \mathbf{T}$ $\qquad\qquad$ (VAR) $\qquad \Gamma \vdash x : \sigma[\tau_i / X_i]$

$$\text{if } x : \sigma = \forall X_1 \cdots X_n . \tau \text{ is in } \Gamma$$
$$\text{and the } \tau_i \text{ are monotypes}$$

(ABS) $\quad \dfrac{\Gamma, x : \tau_1 \vdash M : \tau_2}{\Gamma \vdash \lambda x.M : \tau_1 \to \tau_2}$ $\qquad$ (APP) $\quad \dfrac{\Gamma \vdash M : \tau_1 \to \tau_2 \quad \Gamma \vdash N : \tau_1}{\Gamma \vdash (MN) : \tau_2}$

$(PAIR)$ $\quad \dfrac{\Gamma \vdash M : \tau_1 \quad \Gamma \vdash N : \tau_2}{\Gamma \vdash <M,N> : \tau_1 \times \tau_2}$ $\quad$ $(PROJ)$ $\quad \dfrac{\Gamma \vdash M : \tau_1 \times \tau_2}{\Gamma \vdash \mathrm{p}_i M : \tau_i}$ $\quad i = 1,2$

(LET) $\quad \dfrac{\Gamma \vdash N : \tau_1 \quad \Gamma, x : \forall X_1 \cdots X_n . \tau_1 \vdash M : \tau_2}{\Gamma \vdash (\lambda x.M)N : \tau_2}$

$$\text{where } \{X_1 \cdots X_n\} \text{ is } FTV(\tau_1) - FTV(\Gamma)$$

(GEN) $\quad \dfrac{\Gamma \vdash N : \tau}{\Gamma \vdash N : \forall X_1 \cdots X_n . \tau}$

$$\text{where } \{X_1 \cdots X_n\} \text{ is } FTV(\tau) - FTV(\Gamma)$$

Table 6.1: Type inference rules for an ML-like functional language.

In what follows we will use indifferently *let x = e' in e* or *($\lambda x.e$)e'*, at our best convenience.

In the type-assignment framework, Definition 1.8.1, used to introduce the notion of valid isomorphism, is no longer appropriate. The programs we work with are assigned not only one but several types, and we must take this fact into account. Indeed, the whole point of Definition 1.8.1 was to relate two types A and B when they admit bijective conversion functions; now, in explicitly typed systems, a function that can take an A into a B has exactly the type $A \to B$, but in a type-assignment framework it is no longer so for two reasons:

- A or B can be now quantified types, and in our ML-like systems we do not have types like $A \to B$ in that case

- due to the let typing rule, given a function M with most general type $A \to B$ and an object O with most general type A, the application

(MO) can have, in principle, most general type *strictly more general than B.*

We proceed then as follows.

Definition 6.2.4 *We say that A and B are isomorphic w.r.t. the context Γ $(\Gamma \vdash A \cong B)$ via M, M^{-1} iff using the typing rules given in Definition 6.2.2 the following holds*

- $\forall P, \Gamma \vdash P : A \quad \Rightarrow \quad \Gamma \vdash (MP) : B \quad and \quad \Gamma \vdash M^{-1}(MP) = P : A$

- $\forall Q, \Gamma \vdash Q : B \quad \Rightarrow \quad \Gamma \vdash (M^{-1}Q) : A \quad and \quad \Gamma \vdash M(M^{-1}Q) = Q : B$

Notice that in the empty context all empty types[1] are vacuously isomorphic. For such types the premise of the implication in the definition of isomorphism in the empty context cannot be satisfied, so the implication holds vacuously. This is one reason why $\emptyset \vdash A \cong B$ is *not* a good choice for the notion of isomorphism we need. Furthermore, we look for a notion of a *uniform* isomorphism that does not depend on the context, in the sense that it works in *all* contexts, not just in the empty one. So we are led to the following:

Definition 6.2.5 (ML isomorphism) *We say that A and B are isomorphic $(A \cong B)$ via M, M^{-1} iff $\forall \Gamma, \Gamma \vdash A \cong B$ via M, M^{-1}.*

It is an easy consequence of this definition the fact that M and M^{-1} are invertible, that is to say, $M \circ M^{-1} = \lambda x.x$ and vice-versa, so it is not necessary to require invertibility explicitly.

Remark 6.2.6 (closed terms) *With this definition, the terms M, M^{-1} that can prove an isomorphism $A \cong B$ are closed. This comes from the fact that such terms must work in any context, so cannot have any free variables.*

Now we can easily verify that axiom 12 is in a sense still valid.

Remark 6.2.7 *Let A be $\forall X.\sigma$, where σ is isomorphic to $\mathbf{T}$ via M, M^{-1}. Then it is easy to check that M, M^{-1} provide an ML-isomorphism between $\forall X.\sigma$ and $\mathbf{T}$ also.*

So we must already add to our tentative definition of the Th^{ML} theory the following new axiom **Unit**, that is essentially axiom 12 of $Th^2_{\times \mathbf{T}}$

---

[1] A type A is called *empty* if there is no closed term M of type A, i.e., no closed M s.t. $\emptyset \vdash M : A$.

restricted to ML types. This fact supports our original idea that Th^{ML} is more than just $Th^2_{\times \mathbf{T}}$ less axioms 10, 11 and 12.

(**Unit**) $\forall X.A = \mathbf{T}$ if A is isomorphic to $\mathbf{T}$.

But the real surprise is that we also get a new isomorphism, not derivable in $Th^2_{\times \mathbf{T}}$, that comes out of the peculiar typing rule used to obtain the traditional *let* polymorphism in ML-style languages.

Proposition 6.2.8 *In ML-like languages, the following isomorphism hold*

(**Split**) $\forall X.A \times B \cong \forall X.\forall Y.A \times (B[Y/X])$

Proof. It suffices to provide M and M^{-1} s.t. $\forall \Gamma, \Gamma \vdash A \cong B$ via M, M^{-1}. Let M be $\lambda x.\langle \mathrm{p}_1 x, \mathrm{p}_2 x \rangle$ and M^{-1} be $\lambda x.x$, and let's check the conditions of Definition 6.2.5. Since these are closed terms, the context Γ poses no problem and it is easy to check that, given N s.t. $\Gamma \vdash N : \forall X.A \times B$, we can derive, using the let polymorphic type inference rule, the following typing

$$\Gamma \vdash (\lambda x.\langle \mathrm{p}_1 x, \mathrm{p}_2 x \rangle)N : \forall X.\forall Y.A \times (B[Y/X]).$$

Now, clearly $(\lambda x.x)((\lambda x.\langle \mathrm{p}_1 x, \mathrm{p}_2 x \rangle)N)$ can be assigned type $\forall X.A \times B$.

For the other direction, given N s.t. $\Gamma \vdash N : \forall X.\forall Y.A \times (B[Y/X])$, observe that, by instantiating both X and Y to X, we can derive

$$\Gamma \vdash (\lambda x.x)N : \forall X.A \times B.$$

Now we can use again the let polymorphic inference rule to give back the original type $\forall X.\forall Y.A \times (B[Y/X])$ to the term $(\lambda/x.\langle \mathrm{p}_1 x, \mathrm{p}_2 x \rangle)((\lambda x.x)N)$. $\square$

Notice that there is an implicit side condition on the variable Y: it must not be free in B. Indeed, whenever applied to a type schema, (**Split**) can rename at will the bound type variables occurring in a product type, but it must not identify two different type variables in B.

Example 6.2.9 #let id x = x;;
Value id is <fun> : 'a -> 'a

#let double x = (x,x);;
Value double is <fun> : 'a -> 'a * 'a

#let join = double id;;
Value join is (<fun>,<fun>) : ('a -> 'a) * ('a -> 'a)

```
#(fun (f,g) -> (f,g)) join;;
(<fun>,<fun>) : ('a -> 'a) * ('b -> 'b)
    □
```

Remark 6.2.10 *The isomorphism* **(Split)** *is not derivable in* $Th^2_{\times \mathbf{T}}$.

Indeed, **(Split)** allows to change the number of free type variables even in types that are not isomorphic to the unit type **T**, while all the axioms in $Th^2_{\times \mathbf{T}}$ preserve that number for such types. This fact shows that type-assignment systems allow to prove *strongly* equivalent some proofs that are not so in the explicitly typed versions of the calculus, even at a higher-order. Actually, take the terms that are the natural candidates for proving **(Split)** in $\lambda^2 \beta \eta \pi *$ (they are *retyping functions*, in the terminology of [Mit88]):

$$M = \lambda p : (\forall X.A \times B).\lambda X.\lambda Y.\langle \mathrm{p}_1(p[X]), \mathrm{p}_2(p[Y])\rangle$$

$$M' = \lambda p : (\forall X.\forall Y.A \times B[Y/X]).\lambda X.\langle \mathrm{p}_1(p[X][X]), \mathrm{p}_2(p[X][X])\rangle$$

They provide just a retraction and not an isomorphism: $M' \circ M$ is the identity on $\forall X.A \times B$, but $M \circ M'$ reduces to

$$\lambda z : (\forall X.\forall Y.A \times B[y/x]).\lambda X.\lambda Y.\langle \mathrm{p}_1(z[X][X]), \mathrm{p}_2(z[Y][Y])\rangle$$

which is in normal form, and not the identity.

So the original idea that ML is just a limited version of explicitly typed second-order λ-calculus seems now to be a little less obvious. In (core) ML we cannot do everything we can do in explicitly polymorphic calculi, as it is well known, *but* it is also true that we can do in (core) ML something that cannot be done in the explicitly typed version of second-order λ-calculus. Of course, it is to be noticed that if we take the type-assignment version of the second-order λ-calculus, like the one used by [Kri90], then **(Split)** becomes valid too; the erasure of the normal form of $M \circ M'$ above reduces to the identity with one step of surjective pairing. So it seems that the lesson to be learned here is that we need to be careful when using results from explicitly typed calculi in type-assignment frameworks and vice-versa.

6.3   Completeness and conservativity results

Are there any more unexpected isomorphisms coming out of the *let* construct? What about axioms 10 and 11 of $Th^2_{\times \mathbf{T}}$ we were forced to leave out? Do they induce some other derived isomorphisms on ML types?

Th^{ML}

(1)	$A \times B$	$\begin{array}{c}\lambda x.\langle p_2 x, p_1 x\rangle\\ \cong \\ \lambda x.\langle p_2 x, p_1 x\rangle\end{array}$	$B \times A$
(2)	$A \times (B \times C)$	$\begin{array}{c}\lambda p.\langle\langle p_1 p, p_1 p_2 p\rangle, p_2 p_2 p\rangle\\ \cong \\ \lambda p.\langle p_1 p_1 p, \langle p_2 p_1 p, p_2 p\rangle\rangle\end{array}$	$(A \times B) \times C$
(3)	$(A \times B) \to C$	$\begin{array}{c}\mathbf{curry}\ =\lambda f.\lambda x.\lambda y.(f\langle x,y\rangle)\\ \cong \\ \mathbf{uncurry}\ =\lambda f.\lambda p.(f p_1 p) p_2 p\end{array}$	$A \to (B \to C)$
(4)	$A \to (B \times C)$	$\begin{array}{c}\lambda f.\langle \lambda x.p_1(fx), \lambda x.p_2(fx)\rangle\\ \cong \\ \lambda p.\lambda x.\langle p_1 p x, p_2 p y\rangle\end{array}$	$(A \to B) \times (A \to C)$
(5)	$A \times \mathbf{T}$	$\begin{array}{c}\lambda p.p_1 p\\ \cong \\ \lambda x.\langle x,()\rangle\end{array}$	A
(6)	$A \to \mathbf{T}$	$\begin{array}{c}\lambda f.()\\ \cong \\ \lambda x.\lambda y.()\end{array}$	$\mathbf{T}$
(7)	$\mathbf{T} \to A$	$\begin{array}{c}\lambda f.(f())\\ \cong \\ \lambda x.\lambda y.x\end{array}$	A
(8)	$\forall X.\forall Y.A$	$\begin{array}{c}\lambda x.x\\ \cong \\ \lambda x.x\end{array}$	$\forall Y.\forall X.A$
(9)	$\forall X.A$	$\begin{array}{c}\lambda x.x\\ \cong \\ \lambda x.x\end{array}$	$\forall Y.A[Y/X]$
(Split)	$\forall X.A \times B$	$\begin{array}{c}\lambda x.x\\ \cong \\ \lambda x.\langle p_1 x, p_2 x\rangle\end{array}$	$\forall X.\forall Y.A \times (B[Y/X])$
(Unit)	$\forall X.A$	$\begin{array}{c}f\\ \cong \\ g\end{array}$	$\mathbf{T}$ if $f : A \cong \mathbf{T} : g$

A, B, C arbitrary types, $\mathbf{T}$ a constant for the **Unit** type,
(in axiom 9, X is free for Y in A, $Y \notin \mathrm{FTV}(A)$).
The upper realizer is from left to right, and the lower is from right to left.

Table 6.2: The isomorphisms of types for ML-like languages.

This is not the case, as we will see in a moment, and the axioms we have found sound up to now are also complete, so we can finally give a name to our theory of ML isomorphisms.

Definition 6.3.1 Th^{ML} *is the theory of equality defined by* $Th^2_{\times \mathbf{T}}$ *less axioms 10, 11 and 12 plus* **Unit** *and (***Split***).*

We will present here two main results concerning Th^{ML}: one is completeness for ML isomorphisms, the other is a conservativity result that shows how on ML types Th^{ML} is actually strictly more powerful than $Th^2_{\times \mathbf{T}}$.

Since the details of the proofs are rather technical, we postpone them to the appendix, where the interested reader can find also the necessary technical definitions and references to previous related work. We provide here just the statement of the theorems, with a short sketch of the arguments of the proofs.

6.3.1   Completeness

The theory Th^{ML} can be shown complete by adapting to the type-assignment framework the techniques introduced in [BDCL92]. We first define a *split normal form* (see Definition 6.4.3) of types suggested by the axioms of Th^{ML}, and we notice that completeness for ML isomorphisms reduces to completeness for isomorphisms of *split normal forms*. Then we provide a suitable notion of reduction on ML terms (Definition 6.8.4) that is compatible with type-assignment (Theorem 6.8.5) and allows us to study the invertible terms associated to these latter isomorphisms. We can find a syntactic characterization of such terms (Proposition 6.8.11). This characterization is suitable to show completeness of Th^{ML} by induction on (roughly) the complexity of these invertible terms, but we actually choose to follow another proof strategy. We show a relation holding between type isomorphisms provable by typed and untyped terms (Proposition 6.8.6 and 6.8.11, Corollary 6.8.10) so that we can easily derive the completeness of Th^{ML} from the completeness of $Th^1_{\times \mathbf{T}}$ for the explicitly typed case.

Theorem 6.3.2 *The theory* Th^{ML} *is complete for ML isomorphisms.*

Proof. See Theorem 6.8.13 in Appendix. $\square$

6.3.2   Relating $Th^2_{\times \mathbf{T}}$ and Th^{ML}

As for the relation between $Th^2_{\times \mathbf{T}}$ and Th^{ML}, a careful analysis of the invertible terms in $\lambda^2 \beta \eta \pi *$ allows to show that (**Split**) and **Unit** give us back *the full power* of $Th^2_{\times \mathbf{T}}$ on ML types.

Proposition 6.3.3 *Let A and B be ML types. If $Th^2_{\times \mathbf{T}}$ proves $A = B$, then Th^{ML} proves $A = B$ too.*

Proof. See Theorem 6.9.4 in Appendix. $\square$

To use standard terminology, one could say that on ML types the theory $Th^2_{\times \mathbf{T}}$ is *conservative* over Th^{ML}. Usually, though, when a conservativity result is stated, it refers to some theory Th' that extends a theory Th but does not prove more equations on the language of Th. This is not the case here; since (**Split**) is not derivable in $Th^2_{\times \mathbf{T}}$ (Remark 6.2.10), the theory Th^{ML} is strictly more powerful then $Th^2_{\times \mathbf{T}}$ on ML types. The previous proposition actually states that, on the class of ML types, Th^{ML} is an *extension* of $Th^2_{\times \mathbf{T}}$ and not the reverse, as one could have expected.

⚠ 6.4 Deciding ML isomorphism

The proof of completeness allows to derive an easy decision algorithm for valid isomorphisms of ML types based on a variant of the *narrowing* technique. Every type A is first rewritten to a (unique) type normal form n.f.(A) via a strongly normalizing confluent type rewriting system derived from the axioms of Th^{ML}.[2]

Definition 6.4.1 *(Type rewriting R) Let $>$ be the transitive and substitutive type-reduction relation generated by:*

$$
\begin{array}{ll}
A \times (B \times C) > (A \times B) \times C & \mathbf{T} \times A > A \\
(A \times B) \to C > A \to (B \to C) & A \to \mathbf{T} > \mathbf{T} \\
A \to (B \times C) > (A \to B) \times (A \to C) & \mathbf{T} \to A > A \\
A \times \mathbf{T} > A & \forall X.\mathbf{T} > \mathbf{T}.
\end{array}
$$

Remark 6.4.2 *A type normal form (abbreviated n.f.(A)) of a type A is just a type with the shape $\forall X_1 \cdots X_m.(A_1 \times \cdots \times A_n)$, where no product or unit type appear in the A_i. We call the A_i the coordinates of A.*

Then the presence of (**Split**) suggests a further elaboration up to another normal form (no longer unique).

Definition 6.4.3 *The split-normal-form s.n.f.(A) of a type A is obtained from n.f.(A) by applying as far as possible (**Split**) to rename generic type variables.*

---

[2]The system $>$ is a sub-system of the one we defined in 5.1.1, see Proposition 5.1.2.

Remark 6.4.4 *For any type A, s.n.f.(A) is a type $\forall \vec{X}.(A_1 \times \cdots \times A_n)$, where no product or unit type appears in the A_i and no two A_i share generic type variables.*

In Appendix 6.7, we will show that Th^{ML} proves $A = B$ iff s.n.f.(A) is proven equal to s.n.f.(B) via associativity and commutativity of product, bound variable renaming (axiom 9), quantifier swap (axiom 8) and the derived axiom **Swap**.

This provides us with a naive algorithm to decide equality in Th^{ML}.

Theorem 6.4.5 (decidability of Th^{ML}) *The theory of ML isomorphisms Th^{ML} is decidable.*

Proof. Given types A and B, first reduce them to their split normal forms s.n.f.(A) and s.n.f.(B).

Now, associativity, commutativity of product, **Swap** and quantifier swap do not change the length nor the alphabet of type expressions, so that equality up to these axioms is trivially decidable. Furthermore, associativity and commutativity of product and **Swap** can be applied independently of variable renaming or quantifier swap, **Swap** can be applied independently of associativity and commutativity of product and variable renaming and quantifier swap can be interchanged in a proof, so that any proof of equality

$$s.n.f.(A) = A_1 = \cdots = A_n = s.n.f.(B)$$

of the two split normal forms can be transformed by reordering the axioms used, in a proof

$$s.n.f.(A) = A'_1 = \cdots = A'_n = s.n.f.(B)$$

that uses associativity and commutativity of product, **Swap** and quantifier swap only after the other axioms. We can restrict without loss of generality to proofs having the following shape: a prefix where only variable renaming is used to prove s.n.f.(A) equal to some type expression A', and then a proof of $A' = $ s.n.f.(B) where we have in this order quantifier swap, associativity and commutativity of product and **Swap**.

Unfortunately, variable renaming can change the alphabet of a type expression, so it is potentially dangerous as it generates an infinite class of equal formulae. Thus the prefix of the equality proofs can be of arbitrary length endangering decidability, but it is actually harmless in this context. We are interested only in those proof prefixes that rename s.n.f.(A) to a A' that contain exactly the type variables of s.n.f.(B), as the rest of the proof can't change name of variables, and there is only a finite number of such renamings.

These observations gives us the decision procedure: for every possible variable renaming σ (just $n!$ where n is the number of type variables in s.n.f.(A)), for every possible permutation of the coordinates (there are $n!$ if the coordinates are n), check for equality componentwise up to **Swap** (this requires another factorial step). Succeed if one can find a variable renaming and a permutation of coordinates that provide componentwise equality. Fail otherwise. □

But what about *efficient* decidability? The naive algorithm used to prove decidability seems far from being a practical one. We will describe here an improved decision procedure as it is implemented both in the CAML and CamlLight systems (for code availability, see the Preface).

6.4.1 An improved decision procedure

A careful analysis of the steps performed in the proof of decidability of Th^{ML} can help to develop a much more efficient algorithm that can be used in a practical implementation.

In the CAML system, the availability of a user-level grammar to describe expressions and types of the language makes the first task very easy. A type can be described with the usual concrete syntax; it just suffices to declare it as a type expression (which is recognized by the grammar gtype) to the system by quoting it with <:gtype< >>. Even better, we declare that the standard grammar will be gtype. In this way, we can write our CAML code for the type-rewriting function by cases almost exactly as in Definition 6.4.1. For this reason, the code shown here is the code for the CAML implementation of the search system, that is more readable, even if functionally completely equivalent to that written in CamlLight.

```
#set default grammar gtype:gtype;;

let full_iso = ref true;;

let rec rew_type =
  function
  | <<^x * ^y>>  -> rew_type_irr <<^(rew_type x) * ^(rew_type y)>>
  | <<^x -> ^y>> -> rew_type_irr <<^(rew_type x) -> ^(rew_type y)>>
  |         x    -> x

and rew_type_irr =
  function
    <<^x -> unit>> as t  -> if !full_iso then <<unit>> else t
```

```
| <<unit -> ^x>>         -> x
| <<^x * unit>>          -> x
| <<unit * ^x>>          -> x
| <<(^x * ^y) -> ^z>>  ->
     rew_type_irr <<^x -> ^(rew_type_irr <<^y -> ^z>>)>>
| <<^x -> (^y * ^z)>>  ->
     <<^(rew_type_irr <<^x -> ^y>>) * ^(rew_type_irr <<^x -> ^z>>)>>
| x                      -> x ;;

type Type_Coords == int * gtype list;;

let TypeRewrite t = (flatten (rew_type t))
where rec flatten = function
   <<^x * ^y>> -> let (lgt1,l1) = flatten x and (lgt2,l2) = flatten y
                  in (lgt1+lgt2,append l1 l2)
|       x        -> (1,[x]);;
```

Then we proceed to rename the variables in the coordinates of a type. We also keep track of this renaming, so that later on we can not only say if two types are equal but also up to which renaming of variables.

The function split_vars splits type variables in a type (represented as a coordinate list) by renaming them consecutively starting from start and keeping track of the renamings in a renamings list envlist.

It uses the function shift_compact_vars, which takes a coordinate and returns a copy with the type variables renamed consecutively starting from start and keeping track of the renaming in env.

```
type VarRenaming == (int * int) list;;

let split_vars start ((lgt:int), coords) =
    (let (ren,newcoords) = fold shifter (start, []) coords
     in ren,lgt,newcoords)
    where shifter (start,envlist) coord =
        (let ((ends,env),newcoord) = shift_cpt_vars (start,[]) coord
         in (ends,append env envlist),newcoord)
    where rec shift_cpt_vars (start,env) = function
             <<(^l) ^n>> ->
               let (start_env, l') = fold shift_cpt_vars (start,env) l
               in start_env,<<(^l') ^n>>
           | <<'^i>> ->
               ((start,env),<<'^(assq i env)>> ?
               (start+1,(i,start)::env),<<'^start>>);;

let SplitTR start typ = split_vars start (TypeRewrite typ);;
```

Next, we start by having another, different look at the problem of deciding equality in Th^{ML} to discover that it can usefully be considered a special case of equational unification.

6.4.2   Equality as unification with variable renamings

Since order and name of quantified type variables is irrelevant (axioms 8 and 9), we can consider the problem of deciding

$$\forall \overrightarrow{X}.(A_1 \times \cdots \times A_n) = \forall \overrightarrow{Y}.(B_1 \times \cdots \times B_n)$$

in the last subtheory consisting of axioms 1, 2, 8, 9 and **Swap** as a special case of unification of

$$(A_1 \times \cdots \times A_n) = (B_1 \times \cdots \times B_n),$$

where we are not allowed to substitute arbitrary types for the type variables $\overrightarrow{X}$ and $\overrightarrow{Y}$, but just other type variables with the constraint of not identifying variables that were originally different. Essentially, we restrict to unifiers that are just *bound variable renamings*. We will also call them in the following *consistent* variable renamings.

Again, two split normal forms are equal iff for some permutation σ : $n \to n$ their coordinates A_i and $B_{\sigma(i)}$ can be unified modulo **Swap** with a variable renaming.

Unification up to **Swap** (left-commutativity of $\to$) is decidable (see [Kir85]), so we can perform the necessary unification modulo **Swap** for all permutations and then check if there exists a permutation where unification succeeds with variable renamings.

Divide and Conquer

Actually, since all the variables are distinct in the different components, the result of unification on A_i and $B_{\sigma(i)}$ for a given permutation σ is completely independent of the outcome of unification on the other coordinates: the variable renaming we are looking for is actually made up of n independent variable renamings (one for each coordinate), so we can use a standard quadratic test to check only the $\frac{n(n+1)}{2}$ relevant permutations instead of trying all the $n!$ possible ones.

This is a significant cut-down on the number of coordinates checking; even without adopting dynamic programming techniques, we can see that the complexity goes steeply down from a monstrous $m!n!S$ that corresponds to trying equality modulo **Swap** (of cost S) for all permutations of m variables and all permutation of n coordinates to an average (still fearful, but much lower) $n^2(\frac{m}{n})!S$ that corresponds to testing equality up to **Swap** for each relevant permutation of coordinates and each permutation of the (average) $\frac{m}{n}$ type variables in a coordinate.

But there is still room for improvement.

6.4.3   Dynamic programming

We can now try to attack also the complexity of checking variable-renamings. Instead of the naive approach consisting in, first, generation of all possible variable renaming, and then checking equality up to **Swap**, we can use our knowledge that the needed variable renaming will have to satisfy equality up to **Swap** to significantly cut down the number of renamings to generate and test.

Actually, any variable occurring rightmost in A cannot be moved by left commutativity and must be renamed to a corresponding variable in B occurring in the same position. Any variable in rightmost position provides a part of the renaming that we look for and rules out all the $(n-1)!$ renamings that do not agree.

For example, when trying to show equal

$$A \to B \to X = A' \to B' \to Y,$$

we know that X must be associated with Y, so we need not try renamings that don't do this.

In unification up to left commutativity one takes this fact into account by using suitable flat normal forms [Kir85], where all permutable subformulas are flattened into a list and the only rightmost nonpermutable subformula is singled out. Here we also keep track of the length of the list to improve efficiency.

```
type Head_type == int * gtype;;

type Flat_type = Fl of Head_type * Flat_type list;;

let rec flatten = function
    <<^x -> (^y -> ^z)>> ->
        let nfx = flatten x and nfy = flatten y
        and (Fl ((lgth,exp), chain)) = flatten z
        in Fl ((lgth+2,exp), nfx::nfy::chain)
  | <<^x -> ^y>>         -> Fl ((1,y),[flatten x])
  | <<^y>>               -> Fl ((0,y),[]);;
```

The unification procedure scans this data structure using all the variables in unmovable positions to build partial renamings and stops as soon as an inconsistent variable renaming is reached (for example, as soon as the same variable is forced to be identified to more than one other variable). Anyway, when such inconsistencies are not encountered and when we find variables whose binding is not determined, it is necessary to examine all possible permutations of the flat premises list and the associated renamings, so we need some code to produce permutations.

```
let rec perms = function
   [] -> [[]]
 | x -> list_it (function a -> function z1 ->
                 (list_it (function y -> function z2 -> (a::y)::z2)
                          (perms (except a x)) z1
                 )
                ) x [];;
```

Our algorithm tries to adopt as much as possible dynamaic programming techniques. We keep the current tentative variable renaming, and we fail as soon as it is made inconsistent by variable bindings imposed by the unification procedure.

```
type VarRenaming == (int * int) list;;

let var_renaming = ref ([]:VarRenaming);;
```

A renaming becomes inconsistent as soon as some variable gets bound to more than one other variable, and we check for this event while updating the variable renaming. Notice that, morally, in the case Some bind1, Some bind2 the test bind1==var2 ought to be (bind1==var2) & (bind2==var1) to insure that the new binding is correct, but we rely on the correctness of the preexisting variable renaming (where (var1,var2) occurs $\iff$ (var2,var1) occurs) to be sure that bind1==var2 $\iff$ bind2==var1.

```
let bound var = try (Some(assoc var !var_renaming))
                with failure _ -> None;;

let rename_var (var1,var2) =
   match bound var1, bound var2
   with Some bind1, Some bind2 -> bind1 == var2
   |    None, None ->
           renaming := (var1,var2)::(var2,var1)::!renaming; true
   |    _    -> false;;
```

Then we start to build our unification procedure up to left commutativity on flat normal forms. We proceed as follows to decide if two flat normal forms Fl((n1,mlt1),flat_l1) and Fl((n2,mlt2),flat_l2) are unifiable:

- Check if their flat lists have the same length and fail if it is not the case: **Swap** does not change the length of flat lists.

- **Unify the heads**, which cannot be moved, and build a partial renaming.

- **Unify the flat lists** of premises starting from the partial renaming built during the unification of the heads.

In case of failure, we restore the partial variable renaming to the state before the call to unification in order to allow the backtracking that is necessary to perform unification of the premises lists.

```
exception UNIFYLIST;;

let rec unify_LC (Fl((n1,mlt1),flat_l1)) (Fl((n2,mlt2),flat_l2)) =
    if not (n1 == n2)
    then false
    else let saved_renaming = !renaming
        in if unify_head (mlt1,mlt2) then
            unify_list(flat_l1, flat_l2)
            else (renaming := saved_renaming; false)

and unify_head Heads =
  (exception unifyHead in
  (try unif_rec Heads;true with unifyHead -> false)
    where rec unif_rec = function
     <<'^r>>,<<'^s>> -> if rename_var (r,s)
                          then ()
                          else raise unifyHead
   |      _ ,<<'^x>> -> raise unifyHead
   | <<'^x>>, _       -> raise unifyHead
   | <<(^l1) ^s1>>,<<(^l2) ^s2>>  ->
         if s1==s2 then do_list unif_rec (combine(l1,l2))
                else raise unifyHead
  )


and (unify_list : Flat_type list * Flat_type list -> bool) (a,b) =
    exception found in
    (try do_list
        (function x ->
          let saved_renaming = !renaming in
          unify_map (x, b);
          renaming := saved_renaming; ()
        ) (perms a); false
     with found -> true)
    where rec unify_map = function
        ([],[]) -> raise found
      | (a::resta,b::restb) -> if unify_LC a b
                              then unify_map (resta,restb) else false
      | _ -> raise UNIFYLIST
```

```
in
let unify_Left_Commutative flat1 flat2 =
   renaming:=[];
   let unifiable = unify_LC flat1 flat2 in
   unifiable,!renaming;;
```

Now that we are able to perform unification up to left commutativity, we are almost done. We just need a standard quadratic test to check equality of two lists of coordinates representing a type. This is what is essentially known as *bag equality*. The code that follows implements a specialized version of such test by means of the two functions findiso and quadratic_test.

The function findiso looks in the list of flat type coordinates flat_list for a coordinate unifiable to flat_coords up to left commutativity. In case of success, it returns the renaming produced by the unification and the rest of the list.

```
let rec findiso flat_list flat_coord =
  match flat_list with
     []          ->  (false,[],[])
  | (a::rest) ->
        let (yn,ren) = unify_Left_Commutative flat_coord a
        in if yn  then (true,rest,ren)
           else let (yn,rest2,ren2) = findiso rest flat_coord
                in (yn,a::rest2,ren2);;
```

quadratic_test takes two lists of coordinates and performs a standard quadratic test of equality using findiso to perform a linear search of an element in a list. This function is as curried as possible to take advantage of partial evaluation.

```
let quadratic_test a b =
  let renaming = ref [] in
  let rec q_test =
        function [] ->
           (function [] -> Some !renaming
                    |  _  -> None)
        |   (a::rest_a) as l ->
           (function [] -> None
             | b::rest_b -> let (yn,rest,ren) = findiso l b
                            in if yn
                               then (renaming := append ren !renaming;
                                     q_test rest rest_b)
                               else None)
  in renaming := []; q_test a b;;
```

Now, let's put everything together to get a predicate that tests for equality modulo ML isomorphisms. We try to avoid as much as possible useless work, so in the code for this predicate we reduce the second argument to full flat normal form only after we know that the number of coordinates is the same, and we cannot avoid performing unification, while the first argument is immediately fully reduced in order to take advantage of partial evaluation. The idea is that are_isos will be given a type to build a predicate that tests for equality to the given type.

```
let are_isos a =
  let ((nextvar_a,renaming_a),(lgt_a,typ_coords_a)) = SplitTR 1 a in
  let flat_coords_a = map flatten typ_coords_a
  in function b ->
     let ((_,renaming_b),(lgt_b,typ_coords_b)) = SplitTR nextvar_a b
     in if (lgt_a <> lgt_b)
        then (None,renaming_a,renaming_b)
        else let flat_coords_b = map flatten typ_coords_b
             in  (quadratic_test flat_coords_a flat_coords_b,
                    renaming_a,renaming_b);;
```

Using this function we build a filter to be applied to the CAML system table in search for functions satisfying a type query. We use the renamings built by the rewriting step and the unification step to rename the type of any identifier matching the type query. In this way the user can easily visualize how the retrieved function can be used as he wants.

```
let build_renaming r_unif rena renb =
  let ren_builder r rena (key,val) = (key,assoc (assoc val r) rena)
  in map (ren_builder r_unif (map (function (x,y) -> (y,x)) rena))
         renb;;

let rec rename_type = function
    []  -> (function x -> x)
  | ren -> (function
               <<(^1) ^n>> -> <<^(map (rename_type ren) 1) ^n>>
             | <<'^i>>     -> <<'^(assoc i ren)>>);;

let filter_iso_to a =
    let is_iso_to_a = are_isos a
    in function x ->
      match is_iso_to_a (type_of x) with
        None,_,_ -> ()
      | Some ren_unif, rena_x, renb_y
        -> print (x^" : ");
           if null ren_unif then print_gtype (type_of x)
           else print_gtype
```

```
          (rename_type
            (build_renaming ren_unif rena_x renb_y)
            (type_of x));
        print_newline()
;;
```

We also define a simple parser for strings representing CAML types so that the search functions will appear more user-friendly.

```
let gtype_of_string s =
    match parse_string ("<:gtype<"^s^">>;;") with
    ML E ->
      begin match eval_syntax E with
      dynamic (ty : gtype) -> ty | _ -> failwith "invalid type"
      end
  | _ -> failwith "invalid type";;
```

We now use a simple iterator on the symbol table of the CAML system to build our search command. It will check the type of any defined identifier against the sought one.

```
let search_iso_gen x =
    do_on_variables (filter_iso_to (gtype_of_string x));;

let search_isos x = full_iso := true; search_iso_gen x;;

let search_iso  x = full_iso := false; search_iso_gen x;;
```

Finally, we export the user-level functions: search_isos takes a type and returns all the defined identifiers with types isomorphic to it w.r.t. the full theory of isomorphism, while search_iso is motivated by the following:

Remark 6.4.6 (Isomorphisms without *) *As remarked in [Rit93], the rule* $A \to \mathbf{T} = \mathbf{T}$ *equates too much functions. In ML a function with only does side effects usually returns type* $\mathbf{T}$ *, so that the premise A is relevant to spot the behavior of the function. For this reason, we also provide a* search_iso *routine that returns only the types that can be proved isomorphic without using* $A \to \mathbf{T} = \mathbf{T}$.

6.4.4 Experimental results

The simple code used above to describe the improved search algorithm is actually part of a working implementation for the CAML [CH88] system, whose user-level functional library contains more than 1000 identifiers. No preprocessing of the library is performed, so that the reduction to normal form is repeated for every library identifier on every call to the search procedure. Even if this work could be done once for all, the current performance of the search system is so satisfactory that the additional complications of this preprocessing are not worthwhile.

Here are same examples of usage with their performance. The machine used in the tests is a DECstation 5000 running CAML V3.1.

```
#timers true;;
() : unit

#search_iso "('a ->'b ->'b) * 'b * 'a list -> 'b";;
it_list : ('b -> 'a -> 'b) -> 'b -> 'a list -> 'b
list_it : ('a -> 'b -> 'b) -> 'a list -> 'b -> 'b
() : unit
Evaluation has needed: Runtime: 3.14s GC: 0.96s
```

This first execution time includes the loading time for the search module.

```
#search_iso "('a -> 'b) -> 'a list -> 'b list";;
map : ('a -> 'b) -> 'a list -> 'b list
rev_map : ('a -> 'b) -> 'a list -> 'b list
map_succeed : ('a -> 'b) -> 'a list -> 'b list
() : unit
Evaluation has needed: Runtime: 1.06s GC: 0.91s

#search_iso "int * 'a list -> 'a";;
nth : 'a list -> int -> 'a
item : 'a list -> int -> 'a
() : unit
Evaluation has needed: Runtime: 1.05s GC: 0.88s

#search_iso "string*int -> string";;
first_n_string : int -> string -> string
last_n_string : int -> string -> string
following_word : string -> int -> string
() : unit
Evaluation has needed: Runtime: 1.05s GC: 0.88s

#search_iso "('a -> 'b -> 'c) -> 'a list -> 'b list -> 'c list";;
map2 : ('a -> 'b -> 'c) -> 'a list -> 'b list -> 'c list
() : unit
Evaluation has needed: Runtime: 0.93s GC: 0.88s

#search_iso "('a -> 'b * 'c -> 'a) -> 'a -> 'b list -> 'c list -> 'a";;
it_pair_list : ('a -> 'b * 'c -> 'a) ->
                 'a -> 'b list * 'c list -> 'a
```

```
list_it2 : ('b * 'c -> 'a -> 'a) ->
           'b list -> 'c list -> 'a -> 'a
it_list2 : ('a -> 'b * 'c -> 'a) ->
           'a -> 'b list -> 'c list -> 'a
() : unit
Evaluation has needed: Runtime: 1.01s GC: 0.88s


#search_iso "('a -> 'a) -> 'a list -> 'a list";;
map_share : ('a -> 'a) -> 'a list -> 'a list
share_map_share : ('a -> 'a) -> 'a list -> 'a list
() : unit
Evaluation has needed: Runtime: 1.01s GC: 0.88s


#search_iso "('a * 'b -> 'c) -> 'a -> 'b -> 'c";;
C : ('a -> 'b -> 'c) -> 'b -> 'a -> 'c
uncurry : ('a -> 'b -> 'c) -> 'a * 'b -> 'c
curry : ('a * 'b -> 'c) -> 'a -> 'b -> 'c
() : unit
Evaluation has needed: Runtime: 1.00s GC: 0.88s


#search_iso "bool -> unit";;
trace_standard : bool -> unit
echo_abbrevs : bool -> unit
display_bool : bool -> unit
gc_alarm : bool -> unit
trace_simple : bool -> unit
set_trace : bool -> unit
trace_verbose : bool -> unit
echo : bool -> unit
timers : bool -> unit
set_emphasis : bool -> unit
print_bool : bool -> unit
echo_types : bool -> unit
echo_values : bool -> unit
echo_bool : bool -> unit
timer : bool -> unit
external_address_cache : bool -> unit
verb_system : bool -> unit
set_exhaustive_matches : bool -> unit
() : unit
Evaluation has needed: Runtime: 1.79s GC: 0.89s
```

The CamlLight implementation of the algorithm shows a comparable performance.

6.5   Adding isomorphisms to the ML type-checker

Up to now we have been interested in isomorphisms of types just from the point of view of library searches using types as search keys, that is to say, we used such isomorphisms "after the fact". Such a point of view is not the only interesting one, especially in the case of type-inference languages. One could be tempted to modify the very basic mechanism of the language, the type-inference algorithm, to incorporate these isomorphisms of types in such a way that functions with isomorphic types can be simply interchanged, with the improved algorithm taking care of inserting the correct type-conversion terms where and when necessary.

Now, it is doubtful if the isomorphisms in $Th^1_{\times \mathbf{T}}$ ought to be made part of the type-inference algorithm of an ML-style language essentially for two reasons:

- **Correctness:** the witnesses of the isomorphisms in $Th^1_{\times \mathbf{T}}$ do *change* the original program, so that the intended meaning of the program is not necessarily preserved when the program type-checks up to isomorphisms, but not in the original system. An easy example is the interaction of the commutativity of product on equal types with functions that are not commutative, like subtraction on numbers. There are ways to recover this case (essentially by ruling out commutativity), but the matter is not clear enough to suggest such a modification right now.

- **Complexity:** adding all these isomorphisms at the type-inference level would require unification up to $Th^1_{\times \mathbf{T}}$, which is not known to be decidable (see [NPS93] for recent results), and even equality up to Th^{ML} is at least as hard as Graph Isomorphism (like in the case for $Th^2_{\times \mathbf{T}}$ that we have seen already at the end of Section 5.4; see [Bas90] for details). So such a modification of the ML type-inference algorithm is not clearly feasible.

But if we look closer at (**Split**), we notice that there is something special in it w.r.t. the other isomorphisms. The terms that witness this isomorphism are essentially the identity. The invertible terms associated to *all* the other isomorphisms perform a coding that is simple but does something to the term, while this is not so in the case of $\lambda x.x$ and $\lambda x. < \mathrm{p}_1 x, \mathrm{p}_2 x >$.

Indeed, the only interesting effect of the term $\lambda x. < \mathrm{p}_1 x, \mathrm{p}_2 x >$ is to allow the use of the *let* polymorphism necessary to change the type of the original term. This fact suggests that (**Split**) has more to do with the type-inference algorithm than with the notion of *coding* we found at the basis of the equivalences needed in library searches performed on the basis of the type seen as a search key. So our concern about correctness of the transformation of program induced by the isomorphisms is no longer there if we consider (**Split**) alone. There is *no* transformation of programs, so the intended meaning is surely preserved. We simply type-check more programs, and we will see in a moment that the new program we allow to type-check *should already type-check*. As for complexity, we will propose below a straightforward modification of the type-inference rules that includes (**Split**) at a very reasonable cost.

It is time for a working example. Let's see *the same program* in ML that type-checks only if written "the right way", while with (**Split**) it would type-check in any case. Since it seemingly cracks the ML type-checker, we will call the following program *crack*.

Example 6.5.1   CAML (mips) (V 2-6.1) by INRIA Fri Nov 24 1989

```
#let join = let pair x = (x,x)
            in let id x = x
               in pair id;;
Value join = (<fun>,<fun>) : (('a->'a)*('a->'a))

#let split = let f x = x in (f,f);;
Value split = (<fun>,<fun>) : (('a->'a)*('b->'b))

#let crack f x y = ((fst f) x, (snd f) y);;
Value crack = <fun> : (('a->'b)*('c->'d)->'a->'c->'b*'d)

(* crack on split and different types *)

#crack split 3 true;;
(3,true) : (num * bool)

(* crack on join and different types *)

#crack join 3 true;;

line 1: ill-typed phrase, the constant true of type
bool cannot be used with type instance num in
crack join 3 true
1 error in typechecking
```

```
Typecheck Failed
```
    □

Both functions, `join` and `split`, define a pair of identity functions, but
only the split version survives the test of the context `crack _ 3 true`!

We can try to understand better what is going on by getting rid of the
let construct via the usual translation `let x = e' in e` $\Rightarrow$ `(`$\lambda x.$`e) e'`.

- `join` translates to
  $(\lambda pair.(\lambda f.pair f)(\lambda x.x))(\lambda x. < x, x >)$

- `split` translates to $(\lambda f. < f, f >)(\lambda x.x)$

Now it is easy to see what is going on: `join` and `split` translate to
two terms that are not syntactically equal, but only up to the usual β
conversion. Actually, `join` β-reduces to `split`.

Now, let's recall the key idea in *let* polymorphism. The polymorphic
rule allows to give a type to an application if this application is typable
in the monomorphic system *after one step of evaluation*. That is to say,
to type $(\lambda x.M)N$, we change the type-inference algorithm, that would try
to give a type to $(\lambda x.M)$ and N separately, and only if it succeeds does
it try to type their application. Instead, we look forward *just one* step of
reduction, that is to say, we try to give a type to $M[N/x]$. If we succeed,
that will be the type the original expression $(\lambda x.M)N$ will be given.

Well, `crack split 3 true` is *two* steps from `crack join 3 true`, so
the original form of polymorphic type inference cannot get it! Adding
(**Split**) corresponds in a sense to moving forward more than one step in
the type-inference process.

Remark 6.5.2 *Of course there are lots of terms that are typable in the
monomorphic discipline only after some steps of reductions, but the examples that are usually given typically involve a nontypable subterm that is
erased during these steps of reduction. For example, $(\lambda x.\lambda y.y)\Omega$, where Ω
is a diverging term, is of course not typable, while its reduct $\lambda y.y$ trivially
has a type.*

It is important to notice that this is not the case of `split` *and* `join`,
*as no interesting subterm is erased during the two steps of reductions that
separate them.*

So adding (**Split**) to the type-checker is not just one of the various
possible extensions of ML that can be suggested, but in a sense it is a
necessary completion of a language that allows, as it is now, one way of
defining a pair of identity functions while forbidding another that seems as
perfectly correct.

6.5.1 A modified type-inference algorithm featuring just (Split) polymorphism.

We can easily modify the polymorphic type inference algorithm to accommodate (**Split**) in the type-inference phase. It is just a matter of taking into account the renaming of type variables allowed by this axiom in the polymorphic type inference rule. So it is enough to add to the original ML type-inference algorithm the rule *split-let* of Table 6.3, with priority on the original *let* one. This type-checking algorithm assigns to join the same

The original *let* inference rule

$$(LET) \quad \frac{\Gamma \vdash N : A \quad \Gamma, x : Gen(A) \vdash M : C}{\Gamma \vdash (\lambda x.M)N : C}$$

$Gen(A) = \forall X_1 \cdots X_n.A$ where $\{X_1 \cdots X_n\}$ is $FTV(A) - FTV(\Gamma)$

The *let* inference rule modified as in [DC92]

$$(\text{split} - \text{let}) \quad \frac{\Gamma \vdash N : A \quad \Gamma, x : SplitGen(A) \vdash M : C}{\Gamma \vdash (\lambda x.M)N : C}$$

$SplitGen(A \times B) = \forall X_1 \cdots X_n Y_1 \cdots Y_m.A \times (B[Y_1 \cdots Y_m/X_{i_1} \cdots X_{i_m}])$, where $X_{i_1} \cdots X_{i_m}$ are the type variables shared by A and B, $Y_1 \cdots Y_m$ are fresh type variables.

$SplitGen(A) = Gen(A)$ if A is not a product type.

Table 6.3: Modifying the *let* inference rule: a first attempt.

type as `split`, thus preventing the type error we saw in Example 6.5.1 above.

Adapting an existing type-checker to accommodate this further rule is rather easy. The necessity of checking for shared type variables in product types requires some care in the actual implementation, but there is no need for new, complex unification procedures.

6.5.2 What is special in (Split)

Why is it possible to add seamlessly (**Split**) to the type-checker, while other isomorphisms pose problems? Now, (**Split**) essentially allows to rename the generic variables of any type schema that has a product as the outermost type-constructor in such a way that the two factors of the product do not share any generic type variable, but there are two very crucial properties enjoyed by (**Split**) that make it suitable for use in type-inference:

- it is **identity based:** as pointed out before, its realizer is *the identity*. This means that to convert a given program P from one type to another in the equivalence class of types modulo (**Split**) we need only apply to it a program equivalent to the identity, which does *not* alter P in any way.[3]

- it is an **instantiation isomorphism:** the left-hand side of (**Split**) is a *generic instance*, in the usual sense, of its right-hand side, but not vice-versa. This do provides a best representative in the equivalence classes of the types isomorphic via (**Split**): the most generic type, that is to say the one obtained by applying (**Split**) as far as possible from left to right.

These facts suggest to extend the original ML type-inference algorithm in such a way that the principal type schema inferred for a term is also the most generic one w.r.t. (**Split**), as we have done in the previous section. Anyway, while this extension is obviously sound (it is easy to verify that we preserve all the good properties of the original inference algorithm in [Dam85, Mil78]), it is not so sure that it is the only possible one. Actually, the two properties of (**Split**) that make it a good candidate for extension of the type-checker can be assumed as a criterion. We will use it to select, among the known isomorphisms, which ones are suitable for being incorporated in the type-inference mechanism and which ones are best left to be used in library searches only.

Criterion 6.5.3 (inference-isomorphisms) *Any isomorphism of types that is identity-based (i.e., its realizers are the identity) and that is an instantiation isomorphism (i.e., one side of the isomorphism is a (generic) instance of the other side) can be incorporated in the ML type-inference algorithm.*

For this reason, we will call such isomorphisms *inference-isomorphisms*.

We try now to support this "criterion" and apply it to the isomorphisms known to hold in the case of ML. This will lead us to the discovery of some more inference-isomorphisms than (**Split**), and we will describe in the last section how to modify the existing type-inference algorithm to accommodate the new isomorphisms, both in the pure calculus and in presence of imperative features like reference types and polymorphic exceptions.

---

[3] As is not the case of **curry** and **uncurry**, which do modify the functional behavior of programs.

We think that any inference-isomorphism of types $M : A \cong B : N$ should be incorporated into the type-inference algorithm for ML-like languages and, in our opinion, the following facts strongly support this view.

Correctness and Coherence. If $M : A \cong B : N$, then any program P of type A can be transformed in a program (MP) with type B. Since M is a program equivalent to the identity (as our isomorphism is identity-based), (MP) does exactly what P did, and the program transformation is trivially correct: *inference-isomorphism do not harm*. But B is a type more general than A (since our isomorphism is an instantiation): *inference-isomorphisms improve the system*. Here is an example of such phenomenon, using (**Split**).

Consider the following way of defining a pair of identity functions (syntax of CAML).

```
#let pair x = (x,x);;
Value pair is <fun> : 'a -> 'a * 'a

#let id x = x;;
Value id is <fun> : 'a -> 'a

#let idpair = pair id;;
Value idpair is (<fun>,<fun>) : ('a -> 'a) * ('a -> 'a)
```

This is not the best type one could expect. In fact, applying (**Split**) we can get a better one:

```
#(fun f -> (fst f, snd f)) idpair;;
(<fun>,<fun>) : ('a -> 'a) * ('b -> 'b)
```

Principal Type Schema. If one side of the isomorphism is an instance of the other, then the principal type schema (pts) property of ML-style languages is preserved: we just get a "more general" pts (as in the previous example).

6.5.3 Choosing the right isomorphisms

In Table 6.2 we have an axiomatization of the theory of ML-style isomorphism, together with the realizers associated with each axiom. We can start looking for possible combinations of the axioms that give raise to isomorphisms complying with our criterion. We first notice that axioms 8 and 9 are not interesting to us. Even if they are realized by the identity, neither side is more general than the other. Actually they just tell us that the order and name of generic variables in type schemas are inessential, fact that we already know. Then it is easy to see that the following combinations work.

- axiom 2 (associativity of $\times$), 8 and (**Split**) allow to extend (**Split**) from pairs to n-tuples:

$$\forall X.A_1 \times \cdots \times A_n = \forall X X_2 \cdots X_n.A_1 \times A'_2 \times \cdots \times A'_n \quad (A'_i = A_i[X_i/X])$$

- axiom 4 (distributivity of $\rightarrow$), 8 and (**Split**) allow to extend (**Split**) to higher-order in a controlled way (we assume $X \notin FTV(A_i)$):

$$\forall X.A_1 \rightarrow \cdots \rightarrow A_n \rightarrow (B \times C) = \forall X Y.A_1 \rightarrow \cdots \rightarrow A_n \rightarrow (B \times (C[Y/X]))$$

Now these new inference-isomorphisms can be implemented in the ML type-inference algorithm by smarter and smarter generalization procedures. The first ones require to split all the shared variables in all components of a tuple type, and not just the two components of a product; the second ones require to split the variables of tuples also on the right of an arrow type-constructor, and not only when the product is the toplevel type-constructor (as in (**Split**)). Furthermore, these new isomorphisms can be combined again with themselves allowing more and more splitting of shared generic type variables, and the final picture we get is the following:

Proposition 6.5.4 *Given an ML-type A, the most general type B isomorphic to it via an inference-isomorphism derived by axioms 2, 4 and (**Split**) can be computed inductively as SplitGenIso(A, $\emptyset$), with SplitGenIso(A, V) defined as follows.*

- *SplitGenIso(A $\times$ B, V) = SplitGenIso(A,V) $\times$ SplitGenIso(B',V)* where B' is B with FTV(B) - V replaced by fresh type variables

- *SplitGenIso(A $\rightarrow$ B, V) = A $\rightarrow$ SplitGenIso(B,V $\cup$ FTV(A))*

- *SplitGenIso(C, V) = C* otherwise.

Hence, we propose to extend the ML type-inference mechanism by adding on top of the ordinary generalization mechanism the greater generality provided via inference-isomorphisms by SplitGenIso, as follows.

$$(\textbf{split} - \textbf{gen} - \textbf{let}) \quad \frac{\Gamma \vdash N : A \quad \Gamma, x : SplitGenIso(Gen(A),\emptyset) \vdash M : C}{\Gamma \vdash (\lambda x.M)N : C}$$

This modified rule for typing the *let* construct is the one that has been added to Caml-Light [Ler90, Mau91]. It clearly subsumes the original one, so that all programs accepted by the original ML algorithm can still be typed. But programs like `pair id` above are given more general types, and it is very easy to find programs that type-check only in the new system.

6.5.4 Right isomorphisms in impure context

In the existing implementations of ML, it is necessary to take into account the impure features that are usually supported: references and exceptions. The original ML type-inference algorithm is not sound in the presence of such constructs, as explained in [MT91] pp. 41-46. It is necessary to impose some limitations on the polymorphism that can be attributed to references and exceptions in order to make it sound. There have been several proposals to this extent, that are quite well explained and compared in [Ler92]. Let us just focus here on the solution adopted by SML [MTH90]. Generic variables are divided into imperative (noted '_a) and applicative ones (noted 'a), and a simple restriction on the generalization of imperative variables guarantees soundness. For our isomorphism-based extension, we have a similar restriction. Since an imperative generic variable represents a shared piece of memory, it is unsound to instantiate it to different types, so SplitGenIso(,) must not split imperative generic type variables. A careful analysis shows that this is already guaranteed by the restrictions imposed on imperative variables in SML, combined with the restriction on applicability of (**Split**) on the right of the arrow.

6.6 Conclusion

This chapter has a twofold purpose. On one hand, it completes all the proofs sketched in [DC92], thus providing a firm basis to the theory of ML isomorphisms; on the other, it focuses on the practical issues of library searches, developing in details the search algorithm implemented in the CAML system. This algorithm is more efficient than the ones described in the references and takes full advantage of the new (**Split**) isomorphism to achieve its performance.

The very same new isomorphism (**Split**) leads us to a more consistent version of the original ML type-checker due to Milner [Dam85, DM82, Mil78]. Even if the suggested modification is a minor one, it surely points out how misleading a certain use of terms like *completeness* theorems for type inference algorithms can be: our result does not contradict *completeness* theorems like the one in [Dam85], which states the *completeness* of the inference-algorithm w.r.t the ML type-assignment rules and not the *completeness* of these typing rules w.r.t. to some class of models.

6.7   Some technical Lemmas

This Appendix contains the proofs of Theorems 6.3.2 and Proposition 6.3.3, and the definitions of the technical notions needed for them, in full details.

6.8   Completeness

To show completeness of Th^{ML}, we first notice that each type-reduction rule in $>$ (see Definition 6.4.1) derives from a valid isomorphism. So to each such type reduction is associated an isomorphism, and then, since isomorphisms compose, *any* isomorphism M can be decomposed as in Figure 6.1, where F and G, with their inverses F^{-1} and G^{-1}, are the isomorphisms associated to the rules used to rewrite the types A and B to their *split normal form*.

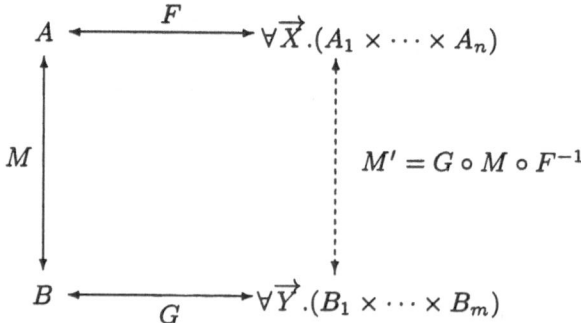

Figure 6.1: Decomposition of an ML isomorphism.

It is evident from the diagram that two types A and B are isomorphic iff their split normal forms are. Now, reduction to split normal form is done accordingly to some axioms of Th^{ML}, so that to prove completeness of this theory it suffices to prove completeness for isomorphisms between types in *split-normal-form*. In order to do this, we study the structure of a generic invertible term providing an isomorphisms between such types. We follow the techniques introduced in [BDCL92] and [DC94] for the case of explicitly typed languages, which we adapt here to the type-assignment framework.

To deal with the structure of terms, we need to work on representatives of the equivalence classes of terms w.r.t. the equality theory of the calculus, such as a normal form. So we first need to provide a suitable notion of reduction that preserves (or at least does not decrease) the set of types that can be assigned to a term. This is not a concern in the case of explicitly typed languages, but in this type-assignment framework it requires some care, as the following remark shows.

Remark 6.8.1 *The reduction rule for surjective pairing*

$$(SP) \qquad \langle p_1 M, p_2 M \rangle \ reduces \ to \ M$$

strictly decreases the set of types that can be assigned to a term by the type-inference algorithm of Definition 6.2.1.

Indeed, consider this simple program which builds a pair of identity functions and then decomposes and builds it up again via projection and pairing.

Example 6.8.2

```
#let splitpair =
#    let join = let pair x = (x,x) in let id x = x in pair id
#in (fst join, snd join);;
Value splitpair is (<fun>,<fun>) : ('a -> 'a) * ('b -> 'b)
```

□

Its most general type is (`'a -> 'a`) * (`'b -> 'b`), and it would reduce, if we allow *SP* contraction, to the following

Example 6.8.3

```
#let splitpair =
#    let join  = let pair x = (x,x) in let id x = x in pair id
#in join;;
Value splitpair is (<fun>,<fun>) : ('a -> 'a) * ('a -> 'a)
```

□

But notice now that (`'a  -> 'a`) * (`'a -> 'a`) cannot be instantiated to (`'a  -> 'a`) * (`'b -> 'b`), which is more general. We lost in the reduction the possibility to instantiate the two components of the product type to two different types.

It is necessary to devise a notion of reduction that is compatible with the type-assignment, that is, that allows us to prove a subject reduction theorem. Actually, if we orient (SP) the other way round to get a surjective pairing *expansion* as described in 2.2.1, we get a weakly normalizing calculus for which the reduct of a term M can be given at least all the types that are legal for M. [4] Notice that since we work in a type-assignment framework, reductions are relativized by a basis Γ where the types of the free term variables are declared.

---

[4]Ongoing research work allows to prove strong normalization for this calculus, but weak normalization is sufficient for the purposes of this chapter, and easier to prove using now standard techniques.

Definition 6.8.4 (notion of reduction for ML)

alpha-beta-eta-csi:

$(\to \beta)$ $\Gamma \vdash (\lambda x.M)N \xrightarrow{ML} M[x:=N] : A$, *if N is free for x in M*

$(\to \eta)$ $\Gamma \vdash M \xrightarrow{ML} \lambda x.(Mx) : A \to B$, *if $x \notin FV(M)$*

surjective pairing:

$(\times \beta)$ *if $\Gamma \vdash \langle M_1, M_2 \rangle : A_1 \times A_2$, $\Gamma \vdash p_i(\langle M_1, M_2 \rangle) \xrightarrow{ML} M_i : A_i$*

$(\times \eta)$ *if $\Gamma \vdash M{:}A \times B$, $\Gamma \vdash M \xrightarrow{ML} \langle p_1(M), p_2(M) \rangle {:} A \times B^5$*

terminal object:

$(*)$ *if $\Gamma \vdash M : \mathbf{T}$ then $\Gamma \vdash M \xrightarrow{ML} *$.*

When talking about reduction, normal form and similar notion in what follows, we will refer to this notion $\xrightarrow{ML}$ of reduction on ML terms.

Theorem 6.8.5 (subject reduction) *Let $\Gamma \vdash M \xrightarrow{ML} M'$. If $\Gamma \vdash M : A$, then $\Gamma \vdash M' : A$.*

Proof. Essentially the same as in [HS80], Theorem 15.17. $\square$

The subject reduction theorem provides us immediately with a nice relation between typings derivable with or without the *let* rule.

Proposition 6.8.6 *Any* closed *term M that is in normal form can be turned into an explicitly typed term of $\lambda^1 \beta \eta \pi *$.*

Proof. Typing with the *let* rule corresponds to typing without the *let* rule after one step of parallel reduction, so, unless there are free variables with polymorphic types, on normal forms the two typing mechanism coincide. Since we do not need polymorphism to type M and since M has no free variables, we can decorate every lambda abstraction with the type in the derivation of $\emptyset \vdash M : A$ and get a term of $\lambda^1 \beta \eta \pi *$ with the same type. $\square$

Remark 6.8.7 *Due to the subject reduction theorem, we can always consider that in $A \cong B$ via M, M^{-1}, the terms M and M^{-1} are in normal form.*

Remark 6.8.8 *Isomorphisms do compose.*

Now we can carry on our analysis of invertible terms. The key results in Section 4.2 of this book can be used to characterize the invertible terms that are typable in ML.

Proposition 6.8.9 *Let $\forall \overrightarrow{X}.(A_1 \times \cdots \times A_n)$ and $\forall \overrightarrow{Y}.(B_1 \times \cdots \times B_m)$ be isomorphic split normal forms in ML. Then $n = m$ and there exists an invertible term M in normal form proving $A \cong B$ and a permutation π : $n \to n$ s.t.*

$$M = \lambda z.\langle M_1[\mathrm{p}_{\pi(1)}z/x_{\pi(1)}], \cdots, M_n[\mathrm{p}_{\pi(n)}z/x_{\pi(n)}]\rangle$$

where $\lambda x_{\pi(i)}.M_i$ are f.h.p.'s.

Proof. It is tedious but straightforward to adapt Proposition 4.2.7 and 4.2.8 to this type-assignment case, with just some care for the (LET) inference rule (when forming an application $(\lambda x.M)N$ one must always do it in the proper order to use the (LET) rule instead of the (APP) rule to avoid restricting too much the types). This is left to the reader. $\square$

Proposition 6.8.6 allows us to prove immediately a nice result on arrow only types.

Corollary 6.8.10 *Let A, B be ML types without occurrence of products or unit type. If $\Gamma \vdash A \cong B$ via invertible terms M, M^{-1} in normal form, then it is possible to decorate M, M^{-1} with types and get terms $M':A' \to B'$, $M'^{-1}:B' \to A'$ of $\lambda^1 \beta \eta \pi *$, where A and B are just $Gen(A')$ and $Gen(B')$ up to generic type variable renaming.*

Proof. By Proposition 6.8.6, any type derivable for M, M^{-1} is derivable without use of *let* (as these terms are closed), so M', M'^{-1} are just the $\lambda^1 \beta \eta \pi *$ terms of Proposition 6.8.6, with types $A' \to B'$ and $B' \to A'$. Furthermore, since these terms are f.h.p.'s, any term variable occurring in them occurs exactly once and bound.

Now recall what $\Gamma \vdash A \cong B$ via M, M^{-1} means: for any term P s.t. $\Gamma \vdash P:A$, $\Gamma \vdash (MP):B$ and $\Gamma \vdash M^{-1}(MP) = P : A$ (and vice-versa).

Even if the *let* rule is not needed in typing M, it can be used in typing the application MP, as M (a f.h.p.) is actually a lambda abstraction. But the bound variable occurs only once in M, so the result of applying the *let* rule is nothing more than an instantiation of A to A' and a renaming and a generalization of B' to B. $\square$

This result can actually be extended to all *split normal forms* by examining the structure of invertible terms that transform *split normal forms* to *split normal forms*.

Proposition 6.8.11 *Let $\forall \overrightarrow{X}.(A_1 \times \cdots \times A_n)$ and $\forall \overrightarrow{Y}.(B_1 \times \cdots \times B_n)$ be isomorphic split normal forms, of length n. Then there exists an invertible term M in normal form proving $A \cong B$ and a permutation $\pi : n \to n$ s.t.*

$$M = \lambda z.\langle M_1[\mathrm{p}_{\pi(1)}z/x_{\pi(1)}], \cdots, M_n[\mathrm{p}_{\pi(n)}z/x_{\pi(n)}]\rangle$$

where $\lambda x_{\pi(i)}.M_i$ are f.h.p.'s that can prove $Gen(A_{\pi(i)})$ isomorphic to a variable renaming of $Gen(B_i)$.

Proof. We first obtain from M, using Proposition 6.8.9 above, the M_i such that $\lambda x_{\pi(i)}.M_i : A' \to B'$, and Corollary 6.8.10 gives the relation between the $A'_i, B'_{\pi(i)}$ and $A_i, B_{\pi(i)}$. $\square$

Remark 6.8.12 *Invertible terms like M in the previous Proposition 6.8.11 are already in normal form (the only potential redexes are the occurrences of z that could be SP expanded, but the expansion is not allowed as z is under the action of a projection).*

The following completeness theorem can now be easily shown using Proposition 4.3.3.

Theorem 6.8.13 (completeness) *The theory Th^{ML} is complete for ML isomorphisms.*

Proof. If two types A and B are isomorphic, then there is an invertible term M proving it of the form shown in Proposition 6.8.11. Then one can show by induction on the Böhm tree of M that $Th^{ML} \vdash A = B$, as in Proposition 4.3.4. $\square$

6.9   Conservativity

Lemma 6.9.1 *Let $M : A \to B$ be a 2-f.h.p. (in normal form). If A and B are types not containing quantifiers, them M is a term of $\lambda^1 \beta \eta$ (the simple typed λ-calculus) and axiom (Swap) suffices to prove $A = B$.*

Proof. By induction on the depth n of the Böhm-tree BT(M) of M.

- $n = 1$.
  Then $M = \lambda x : A.\ x : A \to A$ and $A = B$, so the thesis holds.

- $n = k + 1$.
  Then $M = \lambda z : A.\lambda \overrightarrow{v_i}.z\overrightarrow{P_i}$ where the v_i are all term variables whose type C_i does not contain $\forall$, since B does not contain $\forall$. Now, we know from the definition of second-order f.h.p. that the $\lambda v_i : C_i.P_{\sigma(i)}$ are second-order f.h.p.'s for some permutation $\sigma : n \to n$. Furthermore, we know that the type D_i of each of the P_i has no occurrence of $\forall$ in it since otherwise A would have occurrences of $\forall$, in order for $z\overrightarrow{P_i}$ to type-check.

Summing up, we know that the second-order f.h.p. $\lambda v_i : C_i.P_{\sigma(i)}$ has type $C_i \rightarrow D_{\sigma(i)}$ with no occurrence of $\forall$, and its Böhm-tree has a strictly lower depth than BT(M). So we can apply the induction hypothesis and we get that each of these terms is just a simply typed term, and hence M, which is built up out of them, is a simply typed term.

Secondly, this proves that $Th^1 \vdash C_i = D_{\sigma(i)}$. But

$$A \equiv D_1 \rightarrow \cdots D_n \rightarrow E$$

and

$$B \equiv C_1 \rightarrow \cdots C_n \rightarrow E$$

for some base type E, in order for M to type-check. So we get the second part of the thesis by:

$$
\begin{aligned}
B \quad &\equiv \quad C_1 \rightarrow \cdots C_n \rightarrow E \\
&= \quad D_{\sigma(1)} \rightarrow \cdots D_{\sigma(n)} \rightarrow E \text{ (by the induction hypothesis)} \\
&= \quad D_1 \rightarrow \cdots D_n \rightarrow E \text{ (by swapping the premises)} \\
&\equiv \quad A
\end{aligned}
$$

where the last equality step uses several times the proper axiom of Th^1 in order to rearrange the premises of the $\rightarrow$ in the right order.

$\square$

Theorem 6.9.2 *Let $\forall \vec{X}.A$ and $\forall \vec{Y}.B$ be second-order types such that A and B do not contain quantifiers, products and the unit type. If $Th^2_{\times \mathbf{T}} \vdash \forall \vec{X}.A = \forall \vec{Y}.B$, then $Th^{ML} \vdash \forall \vec{X}.A = \forall \vec{Y}.B$.*

Proof. Suppose that the given types are equal in $Th^2_{\times \mathbf{T}}$. They are already in normal form w.r.t. the rewriting system **R** of Definition 5.1.1, so by Theorem 5.2.17 their isomorphism is witnessed by an invertible term M that is actually a 2-f.h.p. (a term of $\lambda^2 \beta \eta$).

Now, $Th^2_{\times \mathbf{T}}$ does not allow us to change the number of quantifiers in a type unless there is at least an occurrence of the unit type in their scope, and this is forbidden by our hypotheses, so we know that the length n of $\vec{X}$ is equal to that of $\vec{Y}$.

Knowing all this, let's study the term M. It is a 2-f.h.p., so (see Definition 1.9.6)

$$M = \lambda z : (\forall \vec{X}.A).\lambda Y_1 \cdots Y_n.\lambda x_{n+1} \cdots x_{n+k}.z P_1 \cdots P_{n+k}.$$

In a 2-f.h.p., all the abstracted type variables must appear once and only
once at the level immediately below that where they are abstracted, so, due
to the type of z and the fact that A does not contain quantifiers, the first
n P_i's must be exactly the type variables $\overrightarrow{Y}$ in some order. This means
that, for the permutation $\sigma : n + k \to n + k$ associated to the 2-f.h.p. M,
we have that $\lambda x_i.P_{\sigma(i)}$ are 2-f.h.p.'s whose types do not contain quantifiers
(or otherwise, due to the fact that A does not contain quantifiers, M would
not type-check). Hence the real structure of M is

$$\lambda Y_1 \cdots Y_n \lambda x_{n+1} \cdots x_{n+k}.z[Y_{\sigma(1)} \cdots Y_{\sigma(n)}]P_{n+1} \cdots P_{n+k},$$

where we know by Lemma 6.9.1, that the 2-f.h.p.'s $\lambda x_i.P_{\sigma(i)}$ (and hence
the P_{n+i}'s) are simple typed terms of $\lambda^1 \beta \eta$.

Now, by a simple induction on the depth of the Böhm tree of M it is
easy to show that $\forall \overrightarrow{X}.A = \forall \overrightarrow{Y}.B$ can be proved using only (**Swap**) and
axioms 8 and 9, that are all derivable in Th^{ML}. $\square$

Corollary 6.9.3 *Let $\forall \overrightarrow{X}.A$ and $\forall \overrightarrow{Y}.B$ be second-order types as above in
Theorem 6.9.2. Let $\forall \overrightarrow{X'}.A$ and $\forall \overrightarrow{Y'}.B$ be the ML types obtained from them
by erasing all quantifications on type variables not occurring in A and B
respectively. Then $Th^2_{\times \mathbf{T}} \vdash \forall \overrightarrow{X}.A = \forall \overrightarrow{Y}.B$ implies $Th^{ML} \vdash \forall \overrightarrow{X'}.A = \forall \overrightarrow{Y'}.B$*

Proof. Suppose $Th^2_{\times \mathbf{T}} \vdash \forall \overrightarrow{X}.A = \forall \overrightarrow{Y}.B$. The terms P_{n+i}'s and the
variables x_i's in Theorem 6.9.2 contain as free type variables only the $\overrightarrow{Y'}_i$'s,
as only these variables occur in the type B, so we can build the term

$$M' = \lambda w : (\forall \overrightarrow{X'}.A).\lambda \overrightarrow{Y'}.\lambda x_{n+1} \cdots x_{n+k}.w[Y'_\sigma]P_{n+1} \cdots P_{n+k}$$

where Y'_σ is what is left of $Y_{\sigma(1)} \cdots Y_{\sigma(n)}$ after erasing the type variables not
occurring in B.

The term M' type-checks[6] and proves (in $Th^2_{\times \mathbf{T}}$) $\forall \overrightarrow{X'}.A = \forall \overrightarrow{Y'}.B$, so
we can once more apply Theorem 6.9.2 and finally get $Th^{ML} \vdash \forall \overrightarrow{X'}.A =
\forall \overrightarrow{Y'}.B$, as required. $\square$

Theorem 6.9.4 *(Th^{ML} subsumes $Th^2_{\times \mathbf{T}}$ on ML types) Let C and D
be any ML types. If $Th^2_{\times \mathbf{T}} \vdash C = D$, then $Th^{ML} \vdash C = D$.*

---

[6]Notice that $\overrightarrow{Y'}$ and $\overrightarrow{X'}$ have the same length, since the rules in $Th^2_{\times \mathbf{T}}$ do not change
the number of bound variables to prove $\forall \overrightarrow{X}.A = \forall \overrightarrow{Y}.B$.

Proof. Let $C = \forall \overrightarrow{X}.A$ and $C = \forall \overrightarrow{Y}.B$ be ML types equated in $Th^2_{\times \mathbf{T}}$. Take their normal forms n.f.(C) and n.f.(D) w.r.t. the type rewriting system $\mathbf{R}$. We know that, since they are equal in $Th^2_{\times \mathbf{T}}$, there is an n s.t. n.f.(C) $= (C_1 \times \cdots \times C_n)$ and n.f.$(D) = (D_1 \times \cdots \times D_n)$, where no product or unit type appears in the C_i's and the D_i's. Moreover, the rewriting rules in R do not push any $\forall$ inside $\to$ or $\times$, and we start with ML-style types (that have $\forall$ only as the outermost type-constructors), so we know that the C_i and the D_i are still ML-style types. More than that, we know that for some types A_i and B_i not containing quantifiers, $C_i \equiv \forall \overrightarrow{X}.A_i$ and $D_i \equiv \forall \overrightarrow{Y}.B_i$. Now, Theorem 5.2.17 says that there exists a permutation $\sigma : n \to n$ s.t. for all i $Th^2_{\times \mathbf{T}} \vdash \forall \overrightarrow{X}.A_i = \forall \overrightarrow{Y}.B_{\sigma(i)}$. Let's call $\overrightarrow{X_i}$ and $\overrightarrow{Y_i}$ the type variables free in the A_i's and the B_i's, respectively. Now Corollary 6.9.3 states that $Th^{ML} \vdash \forall \overrightarrow{X_i}.A_i = \forall \overrightarrow{Y_{\sigma(i)}}.B_{\sigma(i)}$. Since we can rename bound type variables in Th^{ML}, these equalities can be turned into $Th^{ML} \vdash \forall \overrightarrow{X'_i}.A'_i = \forall \overrightarrow{Y'}_{\sigma(i)}.B'_{\sigma(i)}$ where all the type variables have been renamed in such a way that no two A'_i's or $B'_{\sigma(i)}$'s share any type variable. If M'_i's are the ML terms associated to these equalities in Th^{ML}, then we can build the ML term

$$\lambda w.\langle M'_1(\mathrm{p}_{\sigma(1)} w), \langle \cdots, M'_n(\mathrm{p}_{\sigma(n)} w)\rangle \cdots \rangle$$

that proves

$$Th^{ML} \vdash \forall \overrightarrow{X'_1} \cdots \overrightarrow{X'_n}.(A'_1 \times \cdots \times A'_n) = \forall \overrightarrow{Y'_1} \cdots \overrightarrow{Y'_n}.(B'_1 \times \cdots \times B'_n)$$

These two last types are in normal form w.r.t the type rewriting system $>$, that is a subsystem of $\mathbf{R}$, and moreover all the coordinates have disjoint type variables: they are actually *split normal forms* of C and D.

Now, Th^{ML} proves that any ML type is equal to any of its *split normal forms* (see again Figure 6.1), so, by transitivity, $Th^{ML} \vdash C = D$, as required. $\square$

Remark 6.9.5 *Notice that the proof relies in an essential way on the equivalence between an ML type and its split normal form, which is due to axiom (**Split**). Actually, without it the previous theorem is false, as the following example shows.*

Example 6.9.6
Let A and B be type expressions that cannot be equated in $Th^2_{\times \mathbf{T}}$ nor in Th^{ML}. Then it is easily seen that

$$Th^2_{\times \mathbf{T}} \vdash \forall XY.(X \to (X \to Y) \to A) \times (Y \to (Y \to X) \to B)$$
$$= \forall ZW.(Z \to (Z \to W) \to B) \times (Z \to (Z \to W) \to A).$$

But Th^{ML} without axiom (**Split**) cannot prove it. These types are already in normal form w.r.t. $>$, and there is no way to equate them with only variable renaming, permutation or swapping of premises. $\square$

Chapter 7

Related works,
Future perspectives

In this book we have shown at length how isomorphisms of types can be used either in collaboration with a human user to perform library searches based on types or to improve existing type-checkers. While the only other work that tries to incorporate type isomorphisms at the type-checker level is [Nip90], the library searches issue is probably the most active research topic today. This is clearly related with the rapidly growing size of programs, systems and libraries used in the daily activity by programmers and also by simple users.

In this final chapter, we give a rather comprehensive overview of advanced work related to isomorphisms of types, from more sophisticated search tools based on matching and unification to extensions of the type as search keys paradigm, and we give a perspective of the possible future applications that have not yet been investigated.

⚠ 7.1   Equational matching of types

We will now focus on *matching* types modulo isomorphisms, that is, on deciding if a type A (the *pattern*) can be instantiated via a suitable substitution σ to a type σA that is equal, modulo our theory of isomorphisms, to a given type B (the *subject*).

Why are we interested in such a problem? Because when a user looks for an available function, he has a specific use in mind, while the implementor tries always to provide the most general function in a library. This means that the user might have in mind a type (or functionality) that is *less* general than the ones available in the library.

There are two ways to use matching in order to enhance our search tool to allow the user to perform approximate searches:

- The sought functionality is less general than the available ones and is an instance of the available one: the search tool should then take the user supplied type as the subject and consider the types available in the libraries as patterns. This is the case for example of the function list_it we met in the Introduction; its type is (isomorphic to) ('a -> 'b -> 'b) -> 'a list -> 'b -> 'b, but in many cases it is used at the less general type ('a -> 'a -> 'a) -> 'a list -> 'a -> 'a, and this is the type that the users will probably search for.

- The sought functionality is less general than the available ones, but it is not an instance of the available one. In some cases this is due to the presence of some additional base parameters in the more general library function. The search system could get at it if the user supplies his type as a pattern to be matched against the library types. A typical example of this situation is provided by the sub_string function, whose type is string -> int -> int -> string. This function selects a substring of the first argument, but it can also be used to select an initial substring by fixing the second argument to be the initial position. A user looking for a prefix function of type string -> int -> string, in case he does not find it, could then try a matching search using string ->int -> _x -> string, with _x a matching variable, and find sub_string.

The first of these two motivations is at the base of Mikael Rittri's works on matching modulo $Th^1_{\times T}$. Based on work by Bertrand Lang [Lan78], he showed that matching modulo $Th^1_{\times T}$ is decidable and *finitary* (i.e., there is always a finite set of substitutions that express all possible solutions), and made a first implementation of the matching algorithm that showed its usability in practical library search. In these sections, we will briefly review his results, and hint how they can be extended to the theory Th^{ML}, using previous results presented in this book that can provide a somewhat shorter proof (the reader interested in the fine details of the matching algorithm should refer to the delightful paper [Rit90a] and to Chapter C of [Rit90b]).

Some notations from matching theory

A *matching* problem is a pair $\langle p, s \rangle$ of terms (the pattern and the subject), and a solution to this problem is a substitution σ such that $\sigma p_= s$. An *equational matching* problem is a matching problem with respect to a given equational theory E, that is to say, still a pair $\langle p, s \rangle$ of terms, but a solution to this problem (called a *matcher*) is a substitution σ such that $\sigma p_{=E} s$, where $=_E$ is equality in the theory E.

The theory of associative commutative operators (AC) and that of associative commutative operators with a unit (ACU) are examples of well-known theories for which matching (and also unification) has been studied. See, for example, [FH86, Hul79].

7.1.1 Decomposing the matching problem

Given a problem $\langle p, s \rangle$, we want to develop an algorithm that will find a complete set of matchers modulo Th^{ML}. For this, it is notationally convenient to take into consideration a slightly different notion of normal form for types than the one considered in the previous chapters:

Definition 7.1.1 (stratified regular forms) *A type is in stratified regular form if it is of the form*

$$(SR_1 \to B_1) \times \cdots \times (SR_n \to B_n)$$

where each B_i is not an arrow, a product or a unit type, and each SR_i is in stratified regular form.

Remark 7.1.2 *It is always possible to put a type in stratified regular form using the axioms in Th^{ML}. A way to do this is using the rewriting system $\mathbf{R}$ of Definition 5.1.1 with the rule for currying turned the other way around like in [Rit90a] (such system is still SN and CR).*

Definition 7.1.3 (basis of a stratified regular form) *For any type in stratified regular form $p = (p_1^e \to p_1) \times \cdots \times (p_n^e \to p_n)$ we define the basis of p as $p_1 \times \cdots \times p_n$.*

Our knowledge of the structure of invertible terms allow us to prove easily the following property:

Theorem 7.1.4 (basis of a type) *Any two stratified regular forms of a given type have the same basis (modulo associativity and commutativity of $\times$).*

Proof. Take any stratified regular form of a given type A.

$$B = (B_1^e \to B_1) \times \cdots \times (B_n^e \to B_n).$$

It is easy to observe that the regular form of B (i.e., the normal form w.r.t. the rewriting system $\mathbf{R}$ of Definition 5.1.1) has exactly the same basis, as follows.

First, notice that since the B_i are not products or arrows or $\mathbf{T}$, they are already in normal form. This means that the normal forms of the $B_i^e \to B_i$ are some $B'_{j_1} \to \cdots \to B'_{j_{k_i}} \to B_i$. Then

$$B'_{1_1} \to \cdots \to B'_{1_{k_1}} \to B_1 \times \cdots \times B'_{n_1} \to \cdots \to B'_{n_{k_n}} \to B_n$$

is the normal form of B. Second, recall that A is equal to B and hence isomorphic to it. This means that the regular form of A is also isomorphic to the regular form of B. By means of Theorem 6.4.5, we know that this last isomorphism can be proved by using commutativity and associativity of product and the **Swap** axiom only. These axioms can only shuffle the B_i, so we can conclude that the basis of a stratified regular form of A is a permutation of (i.e., equal modulo associativity and commutativity of product to) the basis of the normal form of A. $\square$

This allows to talk about the basis of a type, even when it is not in stratified regular form. The key of the solution is then the following observation (originally in [Lan78] for a noncommutative version of $Th^1_{\times \mathbf{T}}$) :

Theorem 7.1.5 (decomposition) *If p matches s modulo Th^{ML}, then the basis $p_1 \times \cdots \times p_n$ of p ACU-matches the basis $s_1 \times \cdots \times s_m$ of s.*

Proof. Let $\sigma p =_{Th^{ML}} s$. The property is proved by a simple development of σp, using the fact that the basis of two isomorphic types are AC-equal, proved before. Matching of bases is ACU and not AC as in the case of equality, because a substitution can assign $\mathbf{T}$ to a variable in the basis of the pattern. $\square$

Once this property is proved, the rest follows in a straightforward way. An ACU-matcher σ_{acu} on the basis can be extended to a most general substitution σ s.t. the basis of σp is ACU equal to the basis of s; then we apply the substitution to p and recall the algorithm recursively over the exponents of σp and s. This procedure terminates because the size of the subject in the matching subproblems is strictly smaller than the size of the original matching problem, and only a finite number of matching subproblems is generated each time (this is due to the fact that ACU-matching is *finitary*, and to the fact that each ACU-matcher σ_{acu} on the basis can be extended in an unique way to a most general substitution σ).

Theorem 7.1.6 (matching modulo Th^{ML}) *Matching modulo Th^{ML} is decidable and finitary.*

Proof. See [Rit90a, Rit90b]: the proof there is specialized to $Th^1_{\times \mathbf{T}}$ but extends immediately to Th^{ML} when using Theorem 7.1.5. $\square$

Also, there are some preliminary results on the complexity of the matching algorithm; in [NPS93], it is shown that if the subject is in normal form, then matching is NP-complete.

An implementation of the matching algorithm for LML is available, on request, from Mikael Rittri (e-mail: `rittri@cs.chalmers.se`).

An implementation of the matching algorithm for CamlLight is available, together with a NeXT, Lucid-emacs and emacs-19 interface, from `ftp.inria.fr` in `/lang/caml-light/cl6unix.tar.Z` (look at the files in directory `contrib/search_isos`). This implementation allows the user to perform searches in both of the two ways described above.

7.2   Using equational unification

It would be interesting to be able to combine the two matching search modes seen before into one single search mode, where we allow some parts of the user type A to be more general of the library type B and some others to be less general. This problem, which amounts to find a substitution σ such that $\sigma A = \sigma B$ modulo our theory of isomorphisms, is a special case of equational unification, a problem actively studied in many research areas (see [DJ90] for a nice survey), and has been shown to be undecidable in the case of the theory of isomorphisms for $\lambda^1 \beta\eta\pi*$ in [NPS93], using a reduction to Hilbert's tenth problem. In the same paper, though, it is shown how to perform unification if we give up the two nonlinear axioms of $Th^1_{\times\mathbf{T}}$, which are the following

$$A \to (B \times C) \;=\; (A \to B) \times (A \to C)$$
$$A \to \mathbf{T} \;=\; \mathbf{T}$$

This amounts to restrict the search to the theory of isomorphisms for Symmetric Monoidal Closed Categories [Sol93], that is, to give up the extensionality for the product type. Based on this result, Rittri extended his search tool for LML with this linear unification algorithm, as reported in [Rit93].

It is to be noticed that checking equality modulo this weaker theory is also easier, as shown in a recent work by Soloviev [SE94].

7.3   Extending the paradigm

There are other relevant research works that try to extend the type as search keys paradigm to new languages and/or with richer type systems.

7.3.1   Searching through type classes

In [Mat], Brian Matthews adapted the retrieval methods of Rittri in order to handle the type-class mechanism of Haskell. To handle function overloading, Haskell divides the type universe into a hierarchy of *type classes* and restricts type variables to those classes [WB89]. This type structure adds the additional restriction on retrieval by type that type-matches and unifications must respect the type-class hierarchy. Matthews translates this hierarchy of type classes into an *order-sorted signature* with types as terms over this signature and then uses a modification of the algorithm of [NPS93] to retrieve functions using an order-sorted equational unification algorithm in order to satisfy the extra restriction given above (for order-sorted unification, see for example [AKN86, Kir88, MGS89]). This results in a considerably more focused search in cases where the type classes of a query type

can be given. This work does not use the isomorphism (**Split**).
The retrieval system for Haskell is available from `bmm@inf.rl.ac.uk`.

7.3.2 Searching with more powerful specifications

In a series of papers on the Venari project [RW91, WRZ92], we can find arguments in favor of moving beyond type isomorphism to more powerful formal specifications as search keys. This allows the user to perform a more refined search using more information about the behavior of a function, which is particularly useful in cases where a common type produces many functions in a library. For practicality, such a specification matcher would necessarily use a form of type matching first, as a filter.

The type search tool used by the Venari project is described in [ZW93]. The approach is similar to the others discussed here. One interesting difference is that rather than taking the theory of isomorphisms as a whole and build a search tool out of it, the axioms for currying and associativity and commutativity of the product, together with usual synctactic unification, are used in isolation to build composable matching tools. The user can then pick and compose the tools in the order that he thinks fit to get the right responses to his query.

Notice that this approach is not equivalent to the one described here: not only does ThML include something more (like distributivity and **Split**), but even matching modulo AC and curry is not equivalent to any finite composition of syntactic unification, currying and AC.

The tool is implemented in SML with an interface in Gnu Emacs and available on request from `amy+@cs.cmu.edu`.

7.3.3 Recursive terms and types

It is possible to extend the results of this book rather easily to handle recursive *terms*. Indeed, using the recent results in [DCK94a], it is possible to get a confluent rewriting system for ML with a recursion combinator and then basically all the results in Chapter 6 go through unchanged. Nonetheless, recursion is useless without recursive types, and more work needs to be done in order to deal with recursive types and the sum type-constructor. On the other side, it is rather clear that all algebraic recursive types are isomorphic to the natural numbers, and hence are in the same class of isomorphism. For example, `int` and `unit list` are clearly isomorphic types, with conversion functions

```
let rec int_of_unit_list =
      fun [] -> 0
      | (_::r) -> 1+int_of_unit_list r;;
```

and

```
let rec unit_list_of_int =
        fun 0 -> []
        | n -> ()::(unit_list_of_int (n-1));;
```

But these conversion functions can become then arbitrarily complex: trees can surely be coded as lists, but this is not for free! Are we willing to take all these arbitrarily complex conversion as legal answers to a query? We are very near to a full picture of type isomorphisms for a real programming language and to devise a new refined problem, namely the classification of isomorphic types by the complexity of the encoding functions.

7.3.4   Other applications of type isomorphisms

Finally, let us just hint at another possible application of type isomorphisms. Terms inhabiting isomorphic types are truly interchangeable, as they *do* compute the same class of functions, but they *need not have the same complexity*, so we can expect to exploit these isomorphisms to perform program transformations in the optimizing phase of a compiler, where one has the freedom to choose the most efficient implementation.

7.4   Future work and perspectives

We can say that the use of types modulo type isomorphisms as search keys in functional languages, initially advocated by Rittri and Runciman and Toyn, has by now found a sound theoretical foundation through the results presented in this book and is getting a wide practical acceptance, witnessed by the several implementations already available and used everyday by a growing community, but this is only a starting point. Many other applications can be foreseen.

7.4.1   Design of type systems for functional languages

More recent is the use of type isomorphisms in the design of the type system of programming languages. The essential contribution here is the isolation of the identity-based inference isomorphisms, which allowed us already to extend such a well-established and stable type system as is the type system of core ML. This result is really a promising starting point for future works, which could lead to more flexible type systems for realistic extensions of core ML, taking into account extensible records.

7.4.2 High-level retrieval of objects in object-oriented libraries

If thousands of more-or-less specialized composable objects will really be the future of software, then only a high-level tool able to retrieve objects (whatever they might be) well suited for a specific task will allow us to take advantage of this wealth of functionalities. We can discuss what the tools described here for the core of functional languages are able to suggest. Even if still no clearcut notion of *object* is widely accepted and recognized, an object is nonetheless quite generally understood as something having an internal state (a set of *instance variables*) and a bunch of *methods* for accessing/modifying it. Here the problem is that the *name* of the methods and of the classes can be highly difficult to guess.

If we take as a (partial) formal specification of an object the sum of the partial specifications of its methods (that would correspond exactly to the object type in the objects as records paradigm), then the results presented here apply directly. It is possible to build a retrieval system able to select in an object library those objects having the methods with the correct types with respect to the query. This idea is quite similar to what is proposed in [ZW93] for SML signatures.

Nevertheless, in the presence of state, our results on isomorphic types are no longer guaranteed to be complete. Here is plenty of space for further investigation.

7.4.3 Dynamic composition of software components

A fascinating application of the ideas presented here would be the possibility to dynamically compose different software modules, built in different time by different working groups. Similar to what happens with the it_list example, there are in general many different ways one can design the interface of a module providing the same functionality, and a great amount of time is often spent to write some kind of *adapter* code to convert the existing module interface to the interface our program needs.

More precisely, using a complete theory of isomorphic specifications, one might build a tool able to present all possible ways to combine two or more given modules, let the user choose the one he wants, and automatically generate the adapter's code.

7.4.4 Representation optimization

Finally, we can point out a last field of application. Isomorphisms of types provide a set of possible transformations on data representation that can be exploited at compile time. The key idea is that isomorphic types are not necessarily equivalent in time or space cost, so one can choose the most effective representation for a given execution scheme.

Bibliography

[AB91]   F. Alessi and F. Barbanera. Strong conjunction and intersection types. Dipartimento di Informatica, Universitá di Torino (Italy), manuscript., 1991.

[AEF89]   E. Andureau, P. Enjalbert, and L. Farinas del Cerro. *Logique Temporelle: Sémantique et Validation de programmes parallèles.* Masson, 1989.

[AJ89]   L. Augustsson and T. Johnsson. The chalmers Lazy-ML compiler. *The Computer Journal*, 32(2), 1989.

[Aka93]   Y. Akama. On Mints' reductions for ccc-Calculus. In *Typed Lambda Calculus and Applications*, number 664 in LNCS, pages 1–12. Springer Verlag, 1993.

[AKN86]   H. Aït-Kaci and R. Nasr. A logic programming language with built-in inheritance. *Journal of Logic Programming*, 3:185–215, 1986.

[AL91]   A. Asperti and G. Longo. *Categories, Types, and Structures.* MIT Press, 1991.

[Ame83]   American national standard institute. *The programming language ADA: reference manual*, 1983. ANSI-MIL-STD-1815A-1983.

[Ame91]   P. America. A behavioral approach to subtyping in object-oriented programming languages. In *Proc. of the REX School/Workshop on the Foundation of Object Oriented Languages*, volume 489 of *Lecture Notes in Computer Science*. Springer-Verlag, 1991.

[Bar84]   H. Barendregt. *The Lambda Calculus; Its syntax and Semantics (revised edition).* North Holland, 1984.

[Bas89]   D. Basin. Verification of combinational logic in Nuprl. *Lecture Notes in Computer Science*, 408:333–357, July 1989.

[Bas90]   D. Basin.   Equality of Terms Containing Associative-Commutative Functions and Commutative Binding Operators is Isomorphism Complete in 10th Int. Conf. on Automated Deduction. *Lecture Notes in Computer Science*, 449, July 1990.

[BDCL92] K. Bruce, R. Di Cosmo, and G. Longo.   Provable isomorphisms of types. *Mathematical Structures in Computer Science*, 2(2):231–247, 1992. Proc. of Symposium on Symbolic Computation, ETH, Zurich, March 1990.

[BL85]   K. Bruce and G. Longo. Provable isomorphisms and domain equations in models of typed languages. *ACM Symposium on Theory of Computing (STOC 85)*, May 1985.

[BM77]   J. Bell and M. Machover. *A course in mathematical logic*. North-Holland, 1977.

[BMS80]   R. Burstall, D. MacQueen, and D. Sanella. Hope: An experimental applicative language. In *Proceedings of the LISP Conference*, pages 136–143, Stanford University, Computer Science Department, July 1980.

[BS82]   A. A. Babaev and S. V. Soloviev. Coherence theorem for canonical maps in cartesian closed categories. *Journal of Soviet Mathematics*, 20, 1982.

[BT88]   V. Breazu-Tannen. Combining algebra and higher order types. In IEEE, editor, *Proceedings of the Symposium on Logic in Computer Science (LICS)*, pages 82–90, July 1988.

[BTG94]   V. Breazu-Tannen and J. Gallier. Polymorphic rewiting preserves algebraic confluence. *Information and Computation*, 1994. To appear.

[C+86a]   D. Clement et al. A simple applicative language: Mini-ML. In *Proceedings of the 1986 ACM Conference on LISP and Functional Programming*, pages 13–27. ACM, 1986. Held at MIT, Cambridge, MA.

[C+86b]   R. Constable et al. *Implementing Mathematics with the NUPRL Development System*. Prentice-Hall, 1986.

[CCH90]   P. Canning, W. R. Cook, and W. Hill. Inheritance is not sub-
          typing. *17th Ann. ACM Symp. on Principles of Programming
          Languages (POPL)*, January 1990.

[CDC91]   P.-L. Curien and R. Di Cosmo. A confluent reduction system
          for the λ-calculus with surjective pairing and terminal object.
          In Leach, Monien, and Artalejo, editors, *Intern. Conf. on Au-
          tomata, Languages and Programming (ICALP)*, volume 510 of
          Lecture Notes in Computer Science, pages 291–302. Springer-
          Verlag, 1991.

[CF58]   H. B. Curry and R. Feys. *Combinatory logic I*. Studies in logic
          and the foundations of mathematics. North - Holland, Amster-
          dam, 1958.

[CG90]   P.-L. Curien and G. Ghelli. Confluence and decidability of
          $\beta\eta top_{\leq}$ reduction on $F_{\leq}$. To appear in *Information and Com-
          putation*, 1990.

[CH85]   T. Coquand and G. Huet. Constructions: a higher-order proof
          system for mechanizing mathematics. *EUROCAL85 in LNCS
          203*, 1985.

[CH88]   G. Cousineau and G. Huet. The CAML primer. Technical
          report, LIENS - Ecole Normale Supérieure, 1988.

[CHS72]   H. B. Curry, J. R. Hindley, and J. P. Seldin. *Combinatory logic
          II*. Studies in logic and the foundations of mathematics. North
          - Holland, Amsterdam, 1972.

[Chu32]   A. Church. A set of postulates for the foundation of logic. *An-
          nals of Mathematics*, 33(2):346–366, 1932.

[Col91]   L. Colson. *Représentation intentionnelle d'algorithmes dans les
          systèmes fonctionnels : études de cas*. Thèse de doctorat, Uni-
          versité Paris VII, January 1991.

[Coo92]   W. R. Cook. Interfaces and specifications for the Smalltalk-80
          collection classes. In *OOPSLA '92*, 1992.

[Cub92]   D. Cubric. On free CCC. Distributed on the **types** mailing list,
          1992.

[Dam85]   L. Damas. *Types Disciplines in Programming Languages*. PhD
          thesis, Computer Science Dept., University of Edimburgh, April
          1985.

[DC92]   R. Di Cosmo. Type isomorphisms in a type assignment frame-
           work. In *19th Ann. ACM Symp. on Principles of Programming
           Languages (POPL)*, pages 200–210. ACM, 1992.

[DC93]   R. Di Cosmo. Deciding type isomorphisms in a type assignment
           framework. *Journal of Functional Programming*, 3(3):485–525,
           1993. Special Issue on ML.

[DC94]   R. Di Cosmo. Second order isomorphic types. A proof theo-
           retic study on second order λ-calculus with surjective pairing
           and terminal object. *Information and Computation*, 1994. To
           appear.

[DCK93]   R. Di Cosmo and D. Kesner. A confluent reduction for the
           extensional typed λ–calculus with pairs, sums, recursion and
           terminal object. In A. Lingas, editor, *Intern. Conf. on Au-
           tomata, Languages and Programming (ICALP)*, volume 700 of
           Lecture Notes in Computer Science, pages 645–656. Springer-
           Verlag, 1993.

[DCK94a]   R. Di Cosmo and D. Kesner. Combining first order algebraic
           rewriting systems, recursion and extensional lambda calculi.
           In S. Abiteboul and E. Shamir, editors, *Intern. Conf. on Au-
           tomata, Languages and Programming (ICALP)21*, number 820
           in Lecture Notes in Computer Science, pages 462–472. Springer-
           Verlag, July 1994.

[DCK94b]   R. Di Cosmo and D. Kesner. Simulating expansions without
           expansions. *Mathematical Structures in Computer Science*, 4:1–
           48, 1994. A preliminary version is available as Technical Report
           LIENS-93-11/INRIA 1911.

[DCL91]   R. Di Cosmo and G. Longo. Constuctively equivalent proposi-
           tions and isomorphisms of objects (or terms as natural transfor-
           mations). In *Logic for Computer Science*, volume 21 of *Math-
           ematical Sciences Research Institute Publications*, pages 73–94.
           Springer Verlag, Berkeley, November 1991.

[Dez76]   M. Dezani-Ciancaglini. Characterization of normal forms pos-
           sessing an inverse in the $\lambda\beta\eta$ calculus. *Theoretical Computer
           Science*, 2:323–337, 1976.

[DJ90]   N. Dershowitz and J.-P. Jouannaud. Rewrite systems. In
           J. Van Leeuwen, editor, *Handbook of theoretical computer sci-
           ence*, volume Vol. B : Formal Models and Semantics, chapter 6,
           pages 243–320. The MIT Press, 1990.

[DM82]   L. Damas and R. Milner. Principal type schemes for functional
          programs. In *Ann. ACM Symp. on Principles of Programming
          Languages (POPL)*, pages 207–212. ACM, 1982.

[Dou93]   D. J. Dougherty. Some lambda calculi with categorical sums
          and products. In *Proc. of the Fifth International Conference on
          Rewriting Techniques and Applications (RTA)*, 1993.

[DT69]   J. Doner and A. Tarski. An extended arithmetic of ordinal
          numbers. *Fundamenta Mathematica*, 65:95–127, 1969.

[DVR92]   A. M. Despain and P. Van Roy. High-performance logic pro-
          gramming with the Aquarius prolog compiler. *IEEE COM-
          PUTER*, pages 54–68, January 1992.

[FH86]   F. Fages and G. Huet. Complete sets of unifiers and matchers in
          equational theories. *Theoretical Computer Science*, 43:189–200,
          1986.

[FH88]   A. Field and P. Harrison. *Functional programming*. Addison-
          Wesley, 1988.

[Fre79]   G. Frege.   Begriffsschrift, eine der arithmetischen nachge-
          bildete formelsprache des reinen für das reinene dansken, 1879.
          Reprinted in [vH67], pagg. 1–82.

[Fre93]   G. Frege.   *Grundgesetze der Arithmetik, begriffsschriftlich
          abgeleitet*, volume 1. Pohle, Jena, 1893.

[Fre03]   G. Frege.   *Grundgesetze der Arithmetik, begriffsschriftlich
          abgeleitet*, volume 2. Pohle, Jena, 1903.

[GD80]   G.Huet and D.C.Oppen. Equations and rewrite rules: A sur-
          vey. In R.V.Book, editor, *Formal Language Theory: Perspec-
          tives and Open Problems*. Academic Press, 1980.

[Gir72]   J.-Y. Girard. *Interprétation fonctionelle et élimination des
          coupures dans l'arithmétique d'ordre supérieure*. Thèse de doc-
          torat d'état, Université de Paris VII, 1972.

[GLT90]   J.-Y. Girard, Y. Lafont, and P. Taylor. *Proofs and Types*. Cam-
          bridge University Press, 1990.

[GR83]   A. Goldberg and D. Robson. *Smalltalk-80: The Language and
          its implementation*. Addison-Wesley, 1983.

[Gri83]   D. Gries. *The Science of Programming*. Springer-Verlag, 1983.

[Gur85]   R. Gurevič. Equational theory of positive numbers with expo-
            nentiation. *Proceedings of the American Mathematical Society*,
            94(1):135–141, May 1985.

[Gur90]   R. Gurevič. Equational theory of positive numbers with ex-
            ponentiation is not finitely axiomatizable. *Annals of Pure and
            Applied Logic*, 49:1–30, 1990.

[Har89]   T. Hardin. Confluence results for the pure strong categorical
            logic C.C.L.; λ-calculi as subsystems of C.C.L. *Theoretical Com-
            puter Science*, 65(2):291–342, 1989.

[Hen77]   L. Henkin. The logic of equality. *American Mathematical
            Monthly*, 84:597–612, October 1977.

[Hin69]   R. Hindley. The principal type-scheme of a an object in combi-
            natory logic. *Transactions of the American Mathematical Soci-
            ety*, 146, 1969.

[Hin82]   R. Hindley. The simple semantics for Coppo-Dezani-Sallé types.
            Lecture Notes in Computer Science, 137:212–226, 1982.

[How80]   W. Howard. The formulae-as-types notion of construction. In
            Hindley and Seldin, editors, *To H.B. Curry: Essays in Combi-
            natory Logic, Lambda Calculus and formalism*, pages 479–490.
            Academic Press, 1980.

[How88]   D. Howe. *Automatic reasoning in an Implementation of con-
            structive type theory*. PhD thesis, Cornell University, 1988.

[HPJe92]   P. Hudak, S. Peyton-Jones, and P. W. (editors). Report on the
            programming language Haskell, a non-strict, purely functional
            language (version 1.2). Sigplan Notices, 1992.

[HR84]   C. W. Henson and L. A. Rubel. Some applications of Nevan-
            linna theory to mathematical logic: Identities of exponential
            functions. *Trans. Am. Math. Soc.*, 282(1):1–32, March 1984.

[HS80]   R. Hindley and J. P. Seldin. *Introduction to Combinators and
            λ-calculus*. London Mathematical Society, 1980.

[Hue76]   G. Huet. Résolution d'équations dans les langages d'ordre
            $1, 2, \ldots, \omega$. *Thèse d'Etat, Université Paris VII*, 1976.

[Hue80]   G. Huet. Confluent reductions: abstract properties and applica-
            tions to term rewriting systems. *Journal of the ACM*, 4(27):797–
            821, october 1980.

[Hul79]   J.-M. Hullot. Associative-commutative pattern matching. In *5th Int. Conf. on Artificial Intelligence*, Tokyo, 1979.

[Jay91]   C. B. Jay. Strong normalisation for simply-typed lambda-calculus as in Lambek-Scott. LFCS, University of Edimburgh, February 1991.

[Jay92]   C. B. Jay. Long $\beta\eta$ normal forms and confluence (revised). Technical Report 44, LFCS - University of Edinburgh, August 1992.

[JG92]   C. B. Jay and N. Ghani. The Virtues of Eta-expansion. Technical Report ECS-LFCS-92-243, LFCS, 1992. University of Edimburgh, to appear inJournal of Functional Programming.

[JW72]   K. Jensen and N. Wirth. *PASCAL user manual and report : ISO PASCAL standard.* Springer Verlag, 1972.

[Kir85]   C. Kirchner. *Methodes et utiles de conception systematique d'algoritmes d'unification dans les theories equationnelles.* PhD thesis, Université de Nancy, 1985.

[Kir88]   C. Kirchner. Order-sorted equational unification. In *5thIntern. Conf. on Logic Programming (ICLP)*, Seattle, 1988.

[Klo80]   J. W. Klop. Combinatory reduction systems. *Mathematical Center Tracts*, 27, 1980.

[KP90]   J.-L. Krivine and M. Parigot. Programming with proofs. Technical report, Université de PARIS VII, UFR de Mathématiques, 1990.

[KR35]   S. C. Kleene and J. Rosser. The inconsistency of certain formal theories. *Annals of Mathematics*, 36(2):630–636, 1935.

[KR78]   B. W. Kernighan and D. M. Ritchie. *The C programming language.* Prentice-Hall software series. Prentice Hall, 1978.

[Kri90]   J.-L. Krivine. *Lambda calculus. Types et Modéles.* Masson, 1990.

[L$^+$93]   X. Leroy et al. The Caml Light system, release 0.6. Software and documentation distributed by anonymous FTP on `ftp.inria.fr`, 1993.

[Lan78]   B. Lang. Matching with multiplication and exponentiation. INRIA, Roquencourt, Le Chesnay Cedex 78153, France, May 1978.

[Ler90]   X. Leroy. The ZINC experiment: an economical implementation of the ML language. Technical report 117, INRIA, 1990.

[Ler92]   X. Leroy. *Typage Polymorphe d'un langage algorithmique*. Thèse de doctorat, Université Paris VII, Paris, June 1992.

[Les83]   P. Lescanne. Computer experiments with the REVE term rewriting systems generator. In *Proceedings of 10th ACM Symposium on Principles of Programming Languages*, pages 99–108. Association for Computing Machinery, 1983.

[Les86]   P. Lescanne. REVE a rewrite rule laboratory. In J. Siekmann, editor, *Proc. 8th Conf. on Automated Deduction*, Lecture Notes in Computer Science, pages 696–697, Oxford (England), 1986. Springer Verlag.

[LS86]   J. Lambek and P. J. Scott. *An introduction to higher order categorical logic*. Cambridge University Press, 1986.

[LW93]   X. Leroy and P. Weis. *Manuel de référence du langage Caml*. InterÉditions, 1993.

[Mac81]   A. Macintyre. The laws of exponentiation. In C. Berline, K. McAloon, and J.-P. Ressayre, editors, *Model Theory and Arithmetic*, volume 890 of *Lecture Notes in Mathematics*, pages 185–197. Springer-Verlag, 1981.

[Mac93]   D. B. MacQueen. Reflections on Standard ML. In P. E. Lauer, editor, *Functional Programming, Concurrency, Simulation and Automated Reasoning*, volume 693 of *Lecture Notes in Computer Science*, pages 32–46. Springer-Verlag, 1993.

[Mai90]   H. G. Mairson. Deciding ML typability is complete for deterministic exponential time. In *17th Ann. ACM Symp. on Principles of Programming Languages (POPL)*, pages 382–401, 1990.

[Mar72]   C. F. Martin. Axiomatic bases for equational theories of natural numbers. *Notices of the Am. Math. Soc.*, 19(7):778, 1972.

[Mar92]   S. Martini. Provable isomorphisms, strong equivalence and realizability. In M.-S. et al., editor, *Proceedings of the Fourth Italian Conference on Theoretical Computer Science*, pages 258–268. Word Scientific Publishing Co, 1992.

[Mat]   B. Matthews. Reusing functional code using type classes for library search. Systems Engineering, Rutherford Appleton Laboratory, Didcot, OXON, OX11 0QX, U.K., e-mail: bmm@inf.rl.ac.uk.

[Mau91]   M. Mauny. *Functional programming using Caml Light.* INRIA, 1991. Included in the Caml Light distribution.

[MGS89]   J. Meseguer, J. Goguen, and G. Smolka. Order-sorted unification. *Journal of Symbolic Computation,* 8:383–413, 1989.

[Mil78]   R. Milner. A theory of type polymorphism in programming. *Journal of Computer and System Science,* 17(3):348–375, 1978.

[Min77]   G. Mints. Closed categories and the theory of proofs. *Zapiski Nauchnykh Seminarov Leningradskogo Otdeleniya Matematicheskogo Instituta im. V.A. Steklova AN SSSR,* 68:83–114, 1977.

[Min79]   G. Mints. Teorija categorii i teoria dokazatelstv.I. *Aktualnye problemy logiki i metodologii nauky,* pages 252–278, 1979.

[Mit88]   J. Mitchell. Polymorphic type inference and containment. *Information and Computation,* 76:211–249, 1988.

[ML71]   S. Mac Lane. *Categories for the working mathematician,* volume 5 of *GTM.* Springer, 1971.

[Mor91]   R. Morgan. *Component Library Retrieval using property models.* PhD thesis, University of Durham - England, rick@easby.dur.ac.uk, 1991.

[MT91]   R. Milner and M. Tofte. *Commentary on Standard ML.* The MIT Press, 1991.

[MTH90]   R. Milner, M. Tofte, and R. Harper. *The Definition of Standard ML.* The MIT Press, 1990.

[Nip90]   T. Nipkow. A critical pair lemma for higher-order rewrite systems and its application to λ^*. *First Annual Workshop on Logical Frameworks,* 1990.

[NPS93]   P. Narendran, F. Pfenning, and R. Statman. On the unification problem for Cartesian closed categories. In *Proceedings, Eighth Annual IEEE Symposium on Logic in Computer Science,* pages 57–63, Montreal, Canada, 19–23 June 1993. IEEE Computer Society Press.

[Obt87]   A. Obtulowicz. Algebra of constructions I. The Word Problem
            for Partial Algebras. *Information and Computation*, 73(2):129–
            173, 1987.

[PM89]   C. Paulin-Mohring. Extracting Fω programs from proofs in the
            calculus of construction. In *16th Ann. ACM Symp. on Princi-
            ples of Programming Languages (POPL)*, 1989.

[Pot81]   G. Pottinger. The Church Rosser Theorem for the Typed
            lambda-calculus with Surjective Pairing. *Notre Dame Journal
            of Formal Logic*, 22(3):264–268, 1981.

[Pra71]   D. Prawitz. Ideas and results in proof theory. *Proceedings of
            the 2nd Scandinavian Logic Symposium*, pages 235–307, 1971.

[PRDR94]   A. Piperno and S. Ronchi Della Rocca. Type inference and ex-
            tensionality. In *Proceedings of the Symposium on Logic in Com-
            puter Science (LICS)*, Paris, France, July 1994. IEEE computer
            society Press.

[PV87]   A. Poigné and J. Voss. On the implementation of abstract data
            types by programming language constructs. *Journal of Com-
            puter and System Science*, 34(2-3):340–376, April/June 1987.

[Rey74]   J. Reynolds. Towards a theory of type structure. *Lecture Notes
            in Computer Science*, 19:408–425, 1974.

[Rey84]   J. Reynolds. Polymorphism is not set-theoretic. *Lecture Notes
            in Computer Science*, 173, 1984.

[Rit90a]   M. Rittri. Retrieving library identifiers by equational matching
            of types in 10th Int. Conf. on Automated Deduction. *Lecture
            Notes in Computer Science*, 449, July 1990.

[Rit90b]   M. Rittri. *Searching program libraries by type and proving com-
            piler correctness by bisimulation*. PhD thesis, University of
            Göteborg, Göteborg, Sweden, 1990.

[Rit91]   M. Rittri. Using types as search keys in function libraries. *Jour-
            nal of Functional Programming*, 1(1):71–89, 1991.

[Rit93]   M. Rittri. Retrieving library functions by unifying types mod-
            ulo linear isomorphism. *RAIRO Theoretical Informatics and
            Applications*, 27(6):523–540, 1993.

[Rog88]   H. Rogers, Jr. *Theory of Recursive Functions and Effective Computability.* The MIT Press, Cambridge, Massachusetts; London, England, second edition, 1988.

[RT91]   C. Runciman and I. Toyn. Retrieving re-usable software components by polymorphic type. *Journal of Functional Programming*, 1(2):191–211, 1991.

[RW10]   B. Russel and A. N. Whitehead. *Principia Mathematica.* Cambridge University Press, 1910. Cambridge, England.

[RW91]   E. R. Rollins and J. M. Wing. Specifications as search keys for software libraries. In K. Furukawa, editor, *Eighth International Conference on Logic Programming*, pages 173–187. MIT Press, 91.

[SE94]   S. V. Soloviev and A. E.Andreev. Linear isomorphism of types: a low upper bound for complexity. Technical report, BRICS reports in Computer Science, 1994.

[Sol83]   S. V. Soloviev. The category of finite sets and cartesian closed categories. *Journal of Soviet Mathematics*, 22(3):1387–1400, 1983.

[Sol93]   S. V. Soloviev. A complete axiom system for isomorphism of types in closed categories. In A. Voronkov, editor, *Logic Programming and Automated Reasoning, 4th International Conference*, volume 698 of *Lecture Notes in Artificial Intelligence (subseries of LNCS)*, pages 360–371, St. Petersburg, Russia, 1993. Springer-Verlag.

[Sta83]   R. Statman. λ-definable functionals and $\beta\eta$ conversion. *Arch. Math. Logik*, 23:21–26, 1983.

[Str86]   B. Stroustrup. *The C++ programming language.* Addison-Wesley series in Computer Science. Addison-Wesley, 1986.

[Sza78]   M. E. Szabo. *Algebra of Proofs*, volume 88 of *Studies in logic and the foundations of mathematics.* North - Holland, Amsterdam-New York, 1978.

[Tai67]   W. Tait. Intensional interpretation of functionals of finite type I. *Journal of Symbolic Logic*, 32, 1967.

[Tro86]   A. S. Troelstra. Strong normalization for typed terms with surjective pairing. *Notre Dame Journal of Formal Logic*, 27(4), 1986.

[Tur37]   A. Turing. Computability and λ-definability. *Journal of Symbolic Logic*, 2:153–163, 1937.

[Tur85]   D. A. Turner. Miranda: A non-strict functional language with polymorphic types. In J. P. Jouannaud, editor, *Proceedings of the Conference on Functional Programming Languages and Computer Architecture*, pages 1–16. Springer-Verlag, 1985. Lecture Notes in Computer Science 201.

[vH67]   J. van Heijenoort, editor. *From Frege to Gödel, a source book in mathematical logic*. Harvard University Press, 1967.

[WAL$^+$90] P. Weis, M. V. Aponte, A. Laville, M. Mauny, and A. Suárez. The CAML reference manual. Technical Report 121, INRIA, Roquencourt B.P.105 - 78153 Le Chesnay Cedex - France, September 1990.

[WB89]   P. Wadler and S. Blott. How to make *ad-hoc* polymorphism less ad hoc. In *Proceedings of 16th Annual ACM Symposium on Principles of Programming Languages, Austin, Texas*, 1989.

[Wel]   J. Wells. Typability and type checking in the second-order lambda calculus are equivalent and undecidable. Draft available as `pub/jbw/logic/f-undecidable.ps` from `cs.bu.edu`.

[Wil81]   A. J. Wilkie. On exponentiation – a solution to Tarski's high school algebra problem. Math. Inst. Oxford University (preprint), October 1981.

[WL93]   P. Weis and X. Leroy. *Le langage Caml*. InterÉditions, 1993.

[WRZ92]   J. M. Wing, E. Rollins, and A. M. Zaremski. Thoughts on a Larch/ML and a new application for LP. In *First International Workshop on Larch*, Dedham, MA, July 1992. Also available as CMU-CS-92-135, July 1992.

[ZW93]   A. M. Zaremsky and J. M. Wing. Signature matching: a key to reuse. In *SIGSOFT*, December 1993. Also available as CMU-CS-93-151, May 1993.

Subject index

(ABS), 23
(ABS$_{impl}$), 24
(APPL), 23
(GEN), 25–27, 168
(INST), 25–27
(VAR), 23
(**Split**), 166, 171
Swap, 113, 114
T, 63
β, 22
η, 62
λ-abstraction, 28
λ-calculus, 21
 typed, 23
 untyped, 21, 46
crack, 189
curry, 173
fold_left, 37
foldl, 37
fold, 37, 38
generic, 17, 24
itlist, 37
join, 189
list_it, 37
search_iso, 185
split, 189
uncurry, 173

abstraction, 28
AC matching, 207
ACU matching, 207
ADA, 17, 25
adjunctions, 102

algebra, 43
 Tarski's High School Problem, 43
algebraic
 data types, 63
 rewriting systems, 65, 66
algorithm vs. function, 32
application, 28
arrow, 91
 as implication, 31
 type, 27
associative, 145, 207
associativity, 116, 143, 176, 208, 211
automatic code generation, 32, 213
axiomatization, 45

Böhm tree, 47
bag equality, 183
base of a type, 209
basis of a stratified regular form, 208
bifunctor, 103
bijection, 124–126, 128–130, 137, 139
bijective, 169
browser, 35, 39
 for functional library, 38
 for NEXTSTEP classes, 35

C, 17
calculus
 intuitionistic positive, 45

vs. programming language, 58

CamlLight, 4, 39, 166, 187, 209

canonical
 bijection, 124
  characterization, 124, 137
 form, 83
 representative, 68, 69
 term, 69, 120, 124, 140
  characterization, 124, 137

cartesian product, 42, 102

category, 41
 cartesian, 103
 cartesian closed, 62, 65, 102
 symmetric monoidal, 45
 theory, 41

ccc (Cartesian Closed Category), 42, 43, 45, 102, 103, 115, 221

Church, 22

Church-Rosser Property, 59

classification of software, 12, 34

coercion, 18

combinator, 39, 65, 73

commutative, 145, 207

commutativity, 72, 83, 116, 143, 176, 180, 188, 208, 211

complete
 set of matcher, 207
 theory
  for $\lambda^2\beta\eta\pi$, 145
  for $\lambda^2\beta\eta*$, 145
  for $\lambda^1\beta\eta\pi$, 116
  for $\lambda^1\beta\eta*$, 116
  for regular type, 120

completeness, 44, 48, 49, 102, 113, 115, 116, 120, 123, 125, 139, 140, 168, 174, 175, 196, 200
 of equational theory, 50
 for $\langle N, \uparrow \rangle$, 44
 for $\langle N, 0, + \rangle$, 44

for $\langle N, \uparrow, \cdot \rangle$, 44

for ML, 174, 200

completion, 66, 68, 166

complexity of $Th^2_{\times\mathbf{T}}$, 145

composition
 of λ-terms, 39
 of arrows, 41

Confluence, 59, 64

confluence, 27, 59, 60, 62–65, 67, 70–73, 78, 80–82, 86
 weak, 59

confluent, 60, 63, 65, 67, 69, 70, 72, 76–78, 82, 83, 162, 211

conformance, 35

conjunctive, 121

conservative, 81, 82, 108, 167

conservativity, 64, 67, 81, 82, 172, 174

consistency, 14, 15, 20

consistent, 15, 50, 195

contraction, 62, 63

contractive, 63, 67

contractum, 75

conversion functions, 212

conversion functions, 46

coordinate, 124–126, 128–133, 136–139, 177–179, 183, 184, 203
 of bijection, 125

correctness, 19, 25, 32–34, 115, 181, 189, 193
 partial, 32
 total, 31

criterion, 64, 67, 72, 76, 78

Curien, 3, 63, 66

currification, 116

Curry, 30, 31

Curry-Howard isomorphism, 30

currying, 208, 211

decidability, 65, 67, 81–83, 145,

177
of $Th^2_{\times T}$, 145
of $Th^1_{\times T}$, 115
of Th^{ML}, 176
   improved, 177
of term equality in ccc's, 65
decidable, 102, 115, 125, 145, 167,
   179, 188, 207, 209
decision procedure
   for Th^{ML}
     fast, 177
     simple, 176
decomposition
   of matching, 207
   of isomorphism, 196
deduction rule, 14
derivable, 116, 117, 146, 171, 172,
   175, 198, 199, 202
derivation, 45, 70, 103, 198
Dezani, 49, 113
diamond property, 78
distributed invertibility, 128, 134
   up to substitutions, 134
distributivity of product, 116
dynamic type-checking, 15

entailement, 23
environment, 23
equality
   extensional, 62
   vs. reduction, 59
equation, 5, 33, 40, 43–45, 81, 83,
   102, 105, 175
equational, 50, 62, 64, 65, 81, 83,
   108, 143, 144, 178, 207,
   210, 220, 223
erasure, 48, 113, 114, 172
error, 15, 16, 19, 20, 65, 191
exonent, 209
expansion, 58, 64–66, 197, 200
expansive, 65
exponent, 42, 102

exponential, 26
exponentiation, 43, 44
extension
   conservative, 81
extensional, 62–64, 66, 67
extensionality, 62, 210

factorial, 22, 177
factorization, 73
finite axiomatization, 45
finite hereditary permutation, 47
   second-order, 48
fixpoint, 22
flexibility, 15
formula, 32, 33, 144
formulae, 31, 144, 145, 147, 161
formulae as types paradigm, 31
formulae as types paradigm, 32,
   33
Fortran, 14, 15
Frege, 13, 22
function
   retyping, 172
   vs. set, 41

generalization, 194, 195
Girard, 86

head free, 133
Hilbert, 210
Hindley, 3, 103
Hindley-Rosen Lemma, 67
Howard, 30, 31, 45

identity, 29, 44
implication, 30
inconsistent
   calculus, 22
   renaming, 181
inheritance, 35, 36, 63
instance, 17, 58, 73, 192, 193, 197,
   206, 207, 213
intuitionistic logic, 31, 103, 147

invertible, 39
   characterization
      in $\lambda^2\beta\eta\pi*$, 139
      untyped, 46
   term decomposition , 124
isomorphic type
   first-order, 101
   second-order, 119
isomorphism, 31, 102, 103
   and imperative features, 195
   and invertibility, 46
   componentwise, 116
   Curry-Howard, 31
   decomposition, 124, 196
   definable, 40, 46, 123, 140
   for ML, 166, 170
   identity based, 192
   in categories, 41
   instantiation, 192
   linear, 45
   not definable, 162
   of formulae, 45
   of types, 38
   provable, 49, 53, 64, 103, 114,
      174
   semantic, 39
   theories of, 52
   uniform, 40

Kleene, 22
Klop, 62, 66

Lambek, 66, 82
language
   declarative, 14, 20
   functional, 20
   imperative, 14
LCF, 20
Lisp, 14
LML, 209, 210
logic, 27, 30, 45, 103, 147
Longo, 38

Martin, 44
Martin's identity, 44
matcher, 211
matching, 6, 184, 206, 207, 209,
   210
Matthews, 6, 210
method, 35, 213
Milner, 20, 37
Mints, 64, 66
model, 39, 40, 42, 45, 46, 49, 83,
   102, 103, 115, 143, 146,
   216
modularity of CR and SN, 65
monomorphic, 17, 23, 167, 168,
   190
monotype, 167
morphism, 41

Newman's Lemma, 65
normal form
   flat, 180
   split, 175
normalization, 85
   strategy, 79
   weak, 79
number theory, 43
number theory, 42

object, 12–18, 27, 35, 37, 40–43,
   49, 52, 61, 63, 64, 82,
   102, 115, 116, 129, 169,
   213
object-oriented
   language, 35
   library, 35, 213
Obtulowicz, 83
operational, 61, 62
operator, 19, 22, 44, 207

pairing, 28, 61
Pascal, 16, 17
pattern, 206
pointer, 17–19, 50

safe, 17
    unsafe, 19
polymorphism, 15, 17, 24
    explicit, 25
    implicit, 25, 167
    split, 191, 194
Prawitz, 64, 66
premise, 180–182, 185
product type, 27
program extraction, 32
projection, 28, 61
    arithmetics, 148

quantification, 167, 202
quantifier, 176, 201, 202

realizer, 173, 192, 193
recursion, 23
reducibility, 65, 67
reducibility candidate, 87
    arrow of, 87, 97
    product of, 87, 97
reduction, 28
    step, 58
    strategy, 58
relation, 37, 40, 58, 121, 174, 200
retrieval, 5, 6, 12, 38, 210, 213
REVE, 121
Rittri, 6, 38, 167, 207, 209, 210,
    212
Roger, 3
Runciman, 37, 212
Russel, 13
Russel's paradox, 13

safety, 15
Scott, 82
search tool, 6, 35, 38, 206, 210,
    213
Smalltalk, 35
SML, 37, 38, 211
software, 5, 12, 34, 35, 213
Soloviev, 42, 45, 46

sort
    in ADA, 17
    in C, 17
    in ML, 21
    in Pascal, 16
    in Scheme, 20
sound, 33, 50, 192, 195, 212
soundness, 32, 50, 102, 115, 122,
    168
    of equational theory, 50
SP, 62
specification, 31, 32, 211
    partial, 15, 33, 34
    vs. type, 37
static type-checking, 15
step of reduction, 58
stratified regular form, 208
strong equivalence, 45
subject, 206
subject reduction, 197
substitution, 29, 75, 109, 110, 127,
    128, 131, 133–135, 149,
    154–157, 159, 160, 207,
    209, 210, 215
subtyping, 35, 36, 63
sums, 44, 65
surjective pairing, 102, 111
syntax-directed, 27, 168
System F, 25

Tarski, 43, 44
term, 27, 130
    invertible, 39
    neutral, 86
terminal object, 27, 42, 63, 83,
    116
theories, 43, 49–51, 67, 81, 83,
    116, 145, 207
theory, 5, 12, 13, 29, 31, 40–43,
    45, 50, 53, 60, 62–65, 81,
    102, 103, 108, 113, 115–
    117, 142, 144–146, 166–

168, 170, 174–176, 185,
193, 195, 196, 200, 206,
207, 210, 211, 213
complete, 50
sound, 50
Top, 63, 217
transformation, 61, 189, 193, 213
Troelstra, 82
Turing, 23
type, 15, 27, 33, 63, 193
as search key, 37, 206
assignment, 15
checking, 15, 33, 34
with isomorphism, 188
Church-style, 23
Curry-style, 24
equivalent, 38
for programming, 14
for programming languages,
12, 14
in Mathematical Logic, 13
inclusion, 63
inference, 15, 33, 34
ML, 167
principal schema, 192, 193
recursive, 211
regular, 121
simple, 121
tuple, 194
type assignment
Damas-Milner, 26
syntax directed, 168
type classes, 210
type rewriting, 103, 121, 175
type schema, 167
type-checking
dynamic, 15
static, 15, 19
typing, 15
Damas-Milner, 26
explicit, 15
implicit, 15, 25

strong, 15
syntax-directed, 168
weak, 15

unification, 178, 210
higher-order, 64
order sorted, 211
universal abstraction, 28
universal application, 28

variable
applicative, 195
generic, 21, 195
imperative, 195

Wilkie's identity, 44

Y, 22

Citation Index

Dez76, 46, 112, 125, 130, 135

AB91, 45
AEF89, 32
AJ89, 20
Aka93, 65, 66
AKN86, 210
AL91, 42
Ame83, 17
Ame91, 35

Bar84, 22, 48, 62, 64, 71, 78, 130,
    134
Bas89, 32
Bas90, 145, 188
BDCL92, 27, 52, 68, 81, 131, 174,
    196
BL85, 48, 49, 52, 124, 125, 137
BM77, 31
BMS80, 20
BT88, 73
BTG94, 73

CCH90, 35
CDC91, 27, 60, 63, 65–67
CF58, 13, 62
CG90, 64
CH85, 32
CH88, 20, 35, 166, 185
CHS72, 13
Chu32, 22
Col91, 32
Coo92, 35
Cub92, 65, 66

C^+86a, 168
C^+86b, 32

Dam85, 192, 195
DC92, 38, 191, 195
DC94, 52, 68, 81, 196
DCK93, 65, 66
DCK94a, 65, 66, 211
DCK94b, 65, 66
DCL91, 103
DJ90, 60, 210
DM82, 168, 195
Dou93, 65, 66
DT69, 43
DVR92, 20

FH86, 207
FH88, 20
Fre03, 13
Fre79, 13
Fre93, 13

GD80, 60
Gir72, 24
GLT90, 31, 46, 62, 71, 77, 79, 86,
    95
GR83, 35
Gri83, 32
Gur85, 45
Gur90, 45

Har89, 73, 82
Hen77, 44
Hin69, 167

Hin82, 103
How80, 31
How88, 32
HPJe92, 20
HR84, 45
HS80, 22, 30, 198
Hue76, 64, 66
Hue80, 60
Hul79, 207

Jay92, 65, 66
JG92, 65, 66
JW72, 16

Kir85, 179, 180
Kir88, 210
Klo80, 62, 66
KP90, 32
KR35, 22
KR78, 17
Kri90, 32, 172

Lan78, 207, 209
Ler90, 194
Ler92, 27, 195
Les83, 121
Les86, 121
LS86, 42, 62, 63, 66, 67, 71, 82
LW93, 20, 30
L$^+$93, 20, 30

Mac81, 45
Mac93, 20
Mai90, 26
Mar72, 52
Mar92, 45
Mat, 38, 210
Mau91, 194
MGS89, 210
Mil78, 26, 37, 166, 167, 192, 195
Min77, 42, 64
Min79, 64, 66
Mit88, 172

ML71, 41
Mor91, 38
MT91, 20, 166, 195
MTH90, 20, 27, 166, 195

Nip90, 83, 206
NPS93, 188, 209, 210

Obt87, 83

PM89, 32
Pot81, 62, 66
Pra71, 64, 66
PRDR94, 66
PV87, 63, 66

Rey74, 24
Rit90a, 38, 167, 207–209
Rit90b, 35, 207, 209
Rit91, 37–39, 167, 168
Rit93, 38, 185, 210
Rog88, 23, 44
RT91, 37, 38
RW10, 13
RW91, 38, 211

SE94, 210
Sol83, 42, 52
Sol93, 45, 210
Str86, 19
Sza78, 42

Tai67, 71
Tro86, 62, 82
Tur37, 23
Tur85, 20

vH67, 229

WAL$^+$90, 20
WB89, 210
Wel, 26
Wil81, 44
WL93, 20, 30

WRZ92, 211

ZW93, 38, 211, 213

Progress in Theoretical Computer Science

Progress in Theoretical Computer Science is a series that focuses on the theoretical aspects of computer science and on the logical and mathematical foundations of computer science, as well as the applications of computer theory. It addresses itself to research workers and graduate students in computer and information science departments and research laboratories, as well as to departments of mathematics and electrical engineering where an interest in computer theory is found.

The series publishes research monographs, graduate texts, and polished lectures from seminars and lecture series. We encourage preparation of manuscripts in some form of TeX for delivery in camera-ready copy, which leads to rapid publication, or in electronic form for interfacing with laser printers or typesetters.

Proposals should be sent directly to the Editor, any member of the Editorial Board, or to: Birkhäuser Boston, 675 Massachusetts Ave., Cambridge, MA 02139. The Series includes:

1. Leo Bachmair, *Canonical Equational Proofs*
2. Howard Karloff, *Linear Programming*
3. Ker-I Ko, *Complexity Theory of Real Functions*
4. Guo-Qiang Zhang, *Logic of Domains*
5. Thomas Streicher, *Semantics of Type Theory: Correctness, Completeness and Independence Results*
6. Julian Charles Bradfield, *Verifying Temporal Properties of Systems*
7. Alistair Sinclair, *Algorithms for Random Generation and Counting*
8. Heinrich Hussmann, *Nondeterminism in Algebraic Specifications and Algebraic Programs*
9. Pierre-Louis Curien, *Categorical Combinators, Sequential Algorithms and Functional Programming*
10. J. Köbler, U. Schöning, and J. Torán, *The Graph Isomorphism Problem: Its Structural Complexity*
11. Howard Straubing, *Finite Automata, Formal Logic, and Circuit Complexity*
12. Dario Bini and Victor Pan, *Polynomial and Matrix Computations, Volume 1 Fundamental Algorithms*
13. James S. Royer and John Case, *Subrecursive Programming Systems: Complexity & Succinctness*
14. Roberto Di Cosmo, *Isomorphisms of Types*
15. Erwin Engeler et al., *The Combinatory Programme*